Freshwater Biology

Freshwater Biology

Edited by
Stanley Cain

Larsen & Keller
www.larsen-keller.com

Freshwater Biology
Edited by Stanley Cain
ISBN: 978-1-63549-693-2 (Hardback)

▤ Larsen & Keller

Published by Larsen and Keller Education,
5 Penn Plaza,
19th Floor,
New York, NY 10001, USA

Cataloging-in-Publication Data

Freshwater biology / edited by Stanley Cain.
     p. cm.
Includes bibliographical references and index.
ISBN 978-1-63549-693-2
1. Freshwater biology. 2. Limnology. 3. Aquatic biology. I. Cain, Stanley.
QH96 .F74 2018
551.48--dc23

For more information regarding Larsen and Keller Education and its products, please visit the publisher's website www.larsen-keller.com

Table of Contents

Preface

Limnology as a process refers to the study of ponds, rivers, lakes, wetlands, streams, etc. Freshwater biology is a sub-division of limnology. It is the study of freshwater ecosystems, especially there scientific and biological aspects. It studies in detail the relationship of aquatic plants and animals with their ecosystem along with species distribution. This book is a compilation of chapters that discuss the most vital aspects in the field of freshwater biology. Such selected concepts that redefine this field have been presented in it. For all those who are interested in freshwater biology, this textbook can prove to be an essential guide.

A short introduction to every chapter is written below to provide an overview of the content of the book:

Chapter 1 - Freshwater is the water that occurs naturally on the surface of the Earth. Freshwater rivers are classified as being upland and lowland rivers. The study of aquatic life and ecology found in freshwater bodies is known as freshwater biology. Freshwater biology is best understood in confluence with the major topics listed in the following chapter; **Chapter 2** - The common forms of fresh water are glaciers, icebergs, lakes, groundwater and rivers. Fresh water accounts for 0.003% of all available water on Earth. This chapter is an overview of the subject matter incorporating all the major forms of fresh water; **Chapter 3** - Freshwater plants are classified into Acorus, Alisma, Cabomba and Myriophyllum. Acorus are plants that are found in wetlands and depend on Aerenchyma to supply oxygen to the roots whereas Cabomba are aquatic plants. The major categories of fresh-water plants are dealt with great details in the chapter; **Chapter 4** - Freshwater molluscs live in both flowing water as well as still water. The classes of molluscs can be divided into bivalves, clams, molluscs, etc. Freshwater fish, heron and freshwater crocodile are some of the other topics discussed in this section. This chapter is an overview of the subject matter incorporating all the major categories of freshwater animals.

Finally, I would like to thank my fellow scholars who gave constructive feedback and my family members who supported me at every step.

Editor

An Introduction to Freshwater and Freshwater Biology

Freshwater is the water that occurs naturally on the surface of the Earth. Freshwater rivers are classified as being upland and lowland rivers. The study of aquatic life and ecology found in freshwater bodies is known as freshwater biology. Freshwater biology is best understood in confluence with the major topics listed in the following chapter.

Freshwater

Earth seen from Apollo 17—the Antarctic ice sheet at the bottom of the photograph contains 61% of the fresh water, or 1.7% of the total water, on Earth.

Fresh water is naturally occurring water on Earth's surface in ice sheets, ice caps, glaciers, icebergs, bogs, ponds, lakes, rivers and streams, and underground as groundwater in aquifers and underground streams. Fresh water is generally characterized by having low concentrations of dissolved salts and other total dissolved solids. The term specifically excludes seawater and brackish water although it does include mineral-rich waters such as chalybeate springs. The term "sweet water" (from Spanish "agua dulce") has been used to describe fresh water in contrast to salt water. The term "fresh water" does not have the same meaning as potable water. Much of the surface fresh water and ground water is unsuitable for drinking without some form of purification because of the presence of chemical or biological contaminants.

Systems

Rivers, lakes, and marshlands, such as (from top) South America's Amazon River, Russia's Lake Baikal, and the Everglades in the United States, are types of freshwater systems.

Fresh water habitats are divided into lentic systems, which are the stillwaters including ponds, lakes, swamps and mires; lotic, or running-water systems; and groundwater which flows in rocks and aquifers. There is, in addition, a zone which bridges between groundwater and lotic systems, which is the hyporheic zone, which underlies many larger rivers and can contain substantially more water than is seen in the open channel. It may also be in direct contact with the underlying underground water. The majority of fresh water on Earth is in ice caps.

Sources

The source of almost all fresh water is precipitation from the atmosphere, in the form of mist, rain and snow. Fresh water falling as mist, rain or snow contains materials dissolved from the atmosphere and material from the sea and land over which the rain bearing clouds have traveled. In industrialized areas rain is typically acidic because of dissolved oxides of sulfur and nitrogen formed from burning of fossil fuels in cars, factories, trains and aircraft and from the atmospheric emissions of industry. In some cases this acid rain results in pollution of lakes and rivers.

In coastal areas fresh water may contain significant concentrations of salts derived from the sea if windy conditions have lifted drops of seawater into the rain-bearing clouds. This can give rise to elevated concentrations of sodium, chloride, magnesium and sulfate as well as many other compounds in smaller concentrations.

In desert areas, or areas with impoverished or dusty soils, rain-bearing winds can pick up sand and dust and this can be deposited elsewhere in precipitation and causing the freshwater flow to be measurably contaminated both by insoluble solids but also by the soluble components of those soils. Significant quantities of iron may be transported in this way including the well-documented transfer of iron-rich rainfall falling in Brazil derived from sand-storms in the Sahara in north Africa.

Water Distribution

Water is a critical issue for the survival of all living organisms. Some can use salt water but many organisms including the great majority of higher plants and most mammals must have access to fresh water to live. Some terrestrial mammals, especially desert rodents appear to survive without drinking but they do generate water through the metabolism of cereal seeds and they also have mechanisms to conserve water to the maximum degree.

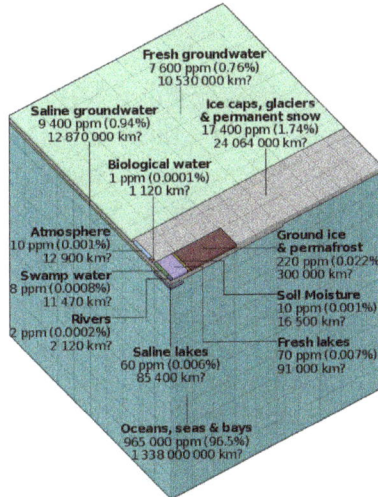

Visualisation of the distribution (by volume) of water on Earth. Each tiny cube (such as the one representing biological water) corresponds to approximately 1000 cubic km of water, with a mass of approximately 1 trillion tonnes (200000 times that of the Great Pyramid of Giza or 5 times that of Lake Kariba, arguably the heaviest man-made object). The entire block comprises 1 million tiny cubes.

Out of all the water on Earth, saline water in oceans, seas and saline groundwater make up about 97% of it. Only 2.5–2.75% is fresh water, including 1.75–2% frozen in glaciers, ice and snow, 0.5–0.75% as fresh groundwater and soil moisture, and less than 0.01% of it as surface water in lakes, swamps and rivers. Freshwater lakes contain about 87% of this fresh surface water, including 29% in the African Great Lakes, 22% in Lake Baikal in Russia, 21% in the North American Great Lakes, and 14% in other lakes. Swamps have most of the balance with only a small amount in rivers, most notably the Amazon River. The atmosphere contains 0.04% water. In areas with no fresh water on the ground surface, fresh water derived from precipitation may, because of its lower density, overlie saline ground water in lenses or layers. Most of the world's fresh water is frozen in ice sheets. Many areas suffer from lack of distribution of fresh water, such as deserts.

Numerical Definition

Fresh water can be defined as water with less than 500 parts per million (ppm) of dissolved salts.

Water salinity based on dissolved salts			
Fresh water	**Brackish water**	**Saline water**	**Brine**
< 0.05%	0.05–3%	3–5%	> 5%

Other sources give higher upper salinity limits for fresh water, e.g. 1000 ppm or 3000 ppm.

Aquatic Organisms

Fresh water creates a hypotonic environment for aquatic organisms. This is problematic for some organisms with pervious skins or with gill membranes, whose cell membranes may burst if excess water is not excreted. Some protists accomplish this using contractile vacuoles, while freshwater fish excrete excess water via the kidney. Although most aquatic organisms have a limited ability to regulate their osmotic balance and therefore can only live within a narrow range of salinity, diadromous fish have the ability to migrate between fresh water and saline water bodies. During these migrations they undergo changes to adapt to the surroundings of the changed salinities; these processes are hormonally controlled. The eel (*Anguilla anguilla*) uses the hormone prolactin, while in salmon (*Salmo salar*) the hormone cortisol plays a key role during this process.

Many sea birds have special glands at the base of the bill through which excess salt is excreted. Similarly the marine iguanas on the Galápagos Islands excrete excess salt through a nasal gland and they sneeze out a very salty excretion.

Fresh Water as a Resource

Water fountain found in a small Swiss village. They are used as a drinking water source for people and cattle. Almost every Alpine village has such a water source.

An important concern for hydrological ecosystems is securing minimum streamflow, especially preserving and restoring instream water allocations. Fresh water is an important natural resource necessary for the survival of all ecosystems. The use of water by humans for activities such as irrigation and industrial applications can have adverse impacts on down-stream ecosystems. Chemical contamination of fresh water can also seriously damage eco-systems.

Pollution from human activity, including oil spills and also presents a problem for freshwater resources. The largest petroleum spill that has ever occurred in fresh water was caused by a Royal Dutch Shell tank ship in Magdalena, Argentina, on 15 January 1999, polluting the environment, drinkable water, plants and animals.

Fresh and unpolluted water accounts for 0.003% of total water available globally.

Agriculture

Changing landscape for the use of agriculture has a great effect on the flow of fresh water. Changes in landscape by the removal of trees and soils changes the flow of fresh water in the local environment and also affects the cycle of fresh water. As a result, more fresh water is stored in the soil which benefits agriculture. However, since agriculture is the human activity that consumes the most fresh water, this can put a severe strain on local freshwater resources resulting in the destruction of local ecosystems. In Australia, over-abstraction of fresh water for intensive irrigation activities has caused 33% of the land area to be at risk of salination. With regards to agriculture, the World Bank targets food production and water management as an increasingly global issue that will foster debate.

Limited Resource

Fresh water is a renewable and variable, but finite natural resource. Fresh water can only be replenished through the process of the water cycle, in which water from seas, lakes, forests, land, rivers, and reservoirs evaporates, forms clouds, and returns as precipitation. Locally however, if more fresh water is consumed through human activities than is naturally restored, this may result in reduced fresh water availability from surface and underground sources and can cause serious damage to surrounding and associated environments.

Fresh Water Withdrawal

Fresh water withdrawal is the quantity of water removed from available sources for use in any purpose, excluding evaporation losses. Water drawn off is not necessarily entirely consumed and some portion may be returned for further use downstream.

Causes of Limited Fresh Water

The increase in the world population and the increase in per capita water use puts increasing strains on the finite resources avialability of clean fresh water The World Bank adds that the response by freshwater ecosystems to a changing climate can be described in terms of three interrelated components: water quality, water quantity or volume, and water timing. A change in one often leads to shifts in the others as well. Water pollution and subsequent eutrophication also reduces the availability of fresh water.

Fresh Water in the Future

Many areas of the world are already experiencing stress on water availability. Due to the accelerated pace of population growth and an increase in the amount of water a single person uses, it is expected that this situation will continue to get worse. A shortage of water in the future would be detrimental to the human population as it would affect everything from sanitation, to overall health and the production of grain.

Choices in the use of Fresh Water

With one in eight people in the world not having access to safe water it is important to use this resource in a prudent manner. Making the best use of water on a local basis probably provides the best solution. Local communities need to plan their use of fresh water and should be made aware of how certain crops and animals use water.

As a guide the following tables provide some indicators.

Table: Recommended basic water requirements for human needs (per person)

Activity	Minimum, litres / day	Range / day
Drinking Water	5	2–5
Sanitation Services	20	20–75
Bathing	15	5–70
Cooking and Kitchen	10	10–50

Table: Water Requirements of different classes of livestock

Animal	Average / day	Range / day
Dairy cow	76 L (20 US gal)	57 to 95 L (15 to 25 US gal)
Cow-calf pair	57 L (15 US gal)	8 to 76 L (2 to 20 US gal)
Yearling cattle	38 L (10 US gal)	23 to 53 L (6 to 14 US gal)
Horse	38 L (10 US gal)	30 to 53 L (8 to 14 US gal)
Sheep	8 L (2 US gal)	8 to 11 L (2 to 3 US gal)

Table: Approximate values of seasonal crop water needs

Crop	Crop water needs mm / total growing period
Banana	1200–2200
Barley/Oats/Wheat	450–650
Cabbage	350–500
Citrus	900–1200
Onions	350–550
Pea	350–500
Potato	500–700
Sugar Cane	1500–2500
Tomato	400–800

Accessing Fresh Water

Canada

Canada has approximately 7% of the world's renewable fresh water. Canadians access their water from ground water, lakes and streams; it is then cleaned and purified in water treatment plants.

United States

The United States uses much more water per capita than developing countries. For example, the average American's daily shower uses more water than a person in a developing country would use for an entire day. Las Vegas, a city that uses a large amount of water to support lush greenery and golf courses, as well as fountains and swimming pools, gets 90% of its water from Lake Mead, which in June 2016 reached its lowest elevation since April 1937.

Developing Countries

In developing countries, 780 million people lack access to clean water. Half of the population of the developing world suffers from at least one disease caused by insufficient water supply and sanitation.

Upland and Lowland

Cascadilla Creek, near Ithaca, New York in the United States, an example of an upland river habitat.

Upland and lowland are conditional descriptions of a plain based on elevation above sea level.

In studies of the ecology of freshwater rivers, habitats are classified as upland and lowland.

Definitions

Upland and lowland are portions of plain that are conditionally categorized by their elevation above the sea level. Lowlands are usually no higher than 200 m (660 ft), while uplands are somewhere around 200 m (660 ft) to 500 m (1,600 ft). On rare occasions, certain lowlands such as Caspian Depression lay below the sea level.

Upland habitats are cold, clear, rocky, fast-flowing rivers in mountainous areas; lowland habitats are warm, slow-flowing rivers found in relatively flat lowland areas, with water that is frequently coloured by sediment and organic matter.

These classifications overlap with the geological definitions of "upland" and "lowland". In geology an "upland" is generally considered to be land that is at a higher elevation than the alluvial plain or stream terrace, which are considered to be "lowlands". The term "bottomland" refers to low-lying alluvial land near a river.

Many freshwater fish and invertebrate communities around the world show a pattern of specialisation into upland or lowland river habitats. Classifying rivers and streams as upland or lowland is important in freshwater ecology as the two types of river habitat are very different, and usually support very different populations of fish and invertebrate species.

Upland

In freshwater ecology, upland rivers and streams are the fast-flowing rivers and streams that drain elevated or mountainous country, often onto broad alluvial plains (where they become lowland rivers). However, altitude is not the sole determinant of whether a river is upland or lowland. Arguably the most important determinants are that of stream power and course gradient. Rivers with a course that drops in altitude rapidly will have faster water flow and higher stream power or "force of water". This in turn produces the other characteristics of an upland river - an incised course, a river bed dominated by bedrock and coarse sediments, a riffle and pool structure and cooler water temperatures. Rivers with a course that drops in altitude very slowly will have slower water flow and lower force. This in turn produces the other characteristics of a lowland river - a meandering course lacking rapids, a river bed dominated by fine sediments and higher water temperatures. Lowland rivers tend to carry more suspended sediment and organic matter as well, but some lowland rivers have periods of high water clarity in seasonal low-flow periods.

The generally clear, cool, fast-flowing waters and bedrock and coarse sediment beds of upland rivers encourage fish species with limited temperature tolerances, high oxygen needs, strong swimming ability and specialised reproductive strategies to prevent eggs or larvae being swept away. These characteristics also encourage invertebrate species with limited temperature tolerances, high oxygen needs and ecologies revolving around coarse sediments and interstices or "gaps" between those coarse sediments.

Lowland

The generally more turbid, warm, slow-flowing waters and fine sediment beds of lowland rivers encourage fish species with broad temperature tolerances and greater tolerances to low oxygen levels, and life history and breeding strategies adapted to these and other traits of lowland rivers. These characteristics also encourage invertebrate species with broad temperature tolerances and greater tolerances to low oxygen levels and ecologies revolving around fine sediments or alternative habitats such as submerged woody debris ("snags") or submergent macrophytes ("water weed").

Amazon River near Manaus, Brazil, an example of a lowland river habitat.

Lowland Alluvial Plains

- American Bottom - flood plain of the Mississippi River in Southern Illinois.

- Bois Brule Bottom

- Bottomland hardwood forest - deciduous hardwood forest found in broad lowland floodplains of the United States.

Freshwater Biology

Freshwater biology is the scientific biological study of freshwater ecosystems and is a branch of limnology. This field seeks to understand the relationships between living organisms in their physical environment. These physical environments may include rivers, lakes, streams, or wetlands. This discipline is also widely used in industrial processes to make use of biological processes such as sewage treatment and water purification. Water flow is an essential aspect to species distribution and influence when and where species interact in freshwater environments.

In the UK the Freshwater Biological Association is based near Windermere in Cumbria.

Freshwater biology is also being used to study the effects of climate change and increased human use.

Freshwater Ecosystem

Freshwater ecosystems are a subset of Earth's aquatic ecosystems. They include lakes and ponds, rivers, streams, springs, and wetlands. They can be contrasted with marine ecosystems, which have a larger salt content. Freshwater habitats can be classified by different factors, including temperature, light penetration, and vegetation.

Freshwater ecosystems can be divided into lentic ecosystems (still water) and lotic ecosystems (flowing water).

Limnology (and its branch freshwater biology) is a study about freshwater ecosystems. It is a part of hydrobiology.

Original attempts to understand and monitor freshwater ecosystems were spurred on by threats to human health (ex. Cholera outbreaks due to sewage contamination). Early monitoring focussed on chemical indicators, then bacteria, and finally algae, fungi and protozoa. A new type of monitoring involves differing groups of organisms (macroinvertebrates, macrophytes and fish) and the stream conditions associated with them.

Current biomonitoring techniques focus mainly on community structure or biochemical oxygen demand. Responses are measured by behavioural changes, altered rates of growth, reproduction or mortality. Macroinvertebrates are most often used in these models because of well known taxonomy, ease of collection, sensitivity to a range of stressors, and their overall value to the ecosystem. Most of these measurements are difficult to extrapolate on a large scale, however.

The use of reference sites is common when assessing what a healthy freshwater ecosystem should "look like". Reference sites are easier to reconstruct in standing water than moving water. Preserved indicators such as diatom valves, macrophyte pollen, insect chitin and fish scales can be used to establish a reference ecosystem representative of a time before large scale human disturbance.

Common chemical stresses on freshwater ecosystem health include acidification, eutrophication and copper and pesticide contamination.

Extinction of Freshwater Fauna

Over 123 freshwater fauna species have gone extinct in North America since 1900. Of North American freshwater species, an estimated 48.5% of mussels, 22.8% of gastropods, 32.7% of crayfishes, 25.9% of amphibians, and 21.3% of fishes are either endangered or threatened. Extinction rates of many species may increase severely into the next century because of invasive species, loss of keystone species, and species which are already functionally extinct. Projected extinction rates for freshwater animals are around five times greater than for land animals, and are comparable to the rates for rainforest communities. Recent extinction trends can be attributed largely to sedimentation, stream fragmentation, chemical and organic pollutants, dams, and invasive species.

Freshwater Ecoregion

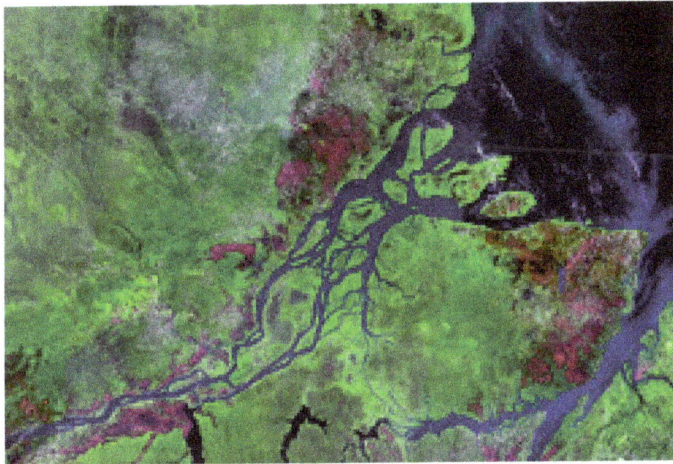

The Amazon River in Brazil.

A freshwater ecoregion is a large area encompassing one or more freshwater systems that contains a distinct assemblage of natural freshwater communities and species. The freshwater species, dynamics, and environmental conditions within a given ecoregion are more similar to each other than to those of surrounding ecoregions and together form a conservation unit. Freshwater systems include rivers, streams, lakes, and wetlands. Freshwater ecoregions are distinct from terrestrial ecoregions, which identify biotic communities of the land, and marine ecoregions, which are biotic communities of the oceans.

A new map of Freshwater Ecoregions of the World, released in 2008, has 426 ecoregions covering virtually the entire non-marine surface of the earth.

World Wildlife Fund (WWF) identifies twelve major habitat types of freshwater ecoregions: Large lakes, large river deltas, polar freshwaters, montane freshwaters, temperate coastal rivers, temperate floodplain rivers and wetlands, temperate upland rivers, tropical and subtropical coastal rivers, tropical and subtropical floodplain rivers and wetlands, tropical and subtropical upland rivers, xeric freshwaters and endorheic basins, and oceanic islands. The freshwater major habitat types reflect groupings of ecoregions with similar biological, chemical, and physical characteristics and are roughly equivalent to biomes for terrestrial systems.

The Global 200, a set of ecoregions identified by WWF whose conservation would achieve the goal of saving a broad diversity of the Earth's ecosystems, includes a number of areas highlighted for their freshwater biodiversity values. The Global 200 preceded Freshwater Ecoregions of the World and incorporated information from regional freshwater ecoregional assessments that had been completed at that time.

References

- "Freshwater". Fishkeeping glossary. Practical Fishkeeping. Archived from the original on 11 May 2006. Retrieved 27 November 2009

- Gordon L., D. M. (2003). "Land cover change and water vapour flows: learning from Australia". Philosophical Transactions of the Royal Society B: Biological Sciences. 358 (1440): 1973–1984. JSTOR 3558315. PMC 1693281. PMID 14728792. doi:10.1098/rstb.2003.1381

- Bisal, G.A.; Specker, J.L. (24 January 2006). "Cortisol stimulates hypo-osmoregulatory ability in Atlantic salmon, Salmo salar L". Journal of Fish biology. Wiley. 39 (3): 421–432. doi:10.1111/j.1095-8649.1991.tb04373.x

- The World Bank, 2009 "Water and Climate Change: Understanding the Risks and Making Climate-Smart Investment Decisions". pp. 19–22. Retrieved 24 October 2011

- Kalujnaia, S.; et. al. (2007). "Salinity adaptation and gene profiling analysis in the European eel (Anguilla anguilla) using microarray technology". Gen Comp Endocrinol. National Center for Biotechnology Information. 152 (2007): 274–80. PMID 17324422. doi:10.1016/j.ygcen.2006.12.025

- Reengaging in Agricultural Water Management: Challenges and Options, The World Bank, pp. 4–5, retrieved 30 October 2011

2

Various Forms of Freshwater

The common forms of fresh water are glaciers, icebergs, lakes, groundwater and rivers. Fresh water accounts for 0.003% of all available water on Earth. This chapter is an overview of the subject matter incorporating all the major forms of fresh water.

Glacier

The Baltoro Glacier in the Karakoram mountains of Pakistan. At 62 kilometres (39 miles) in length, it is one of the longest alpine glaciers on earth.

A glacier is a persistent body of dense ice that is constantly moving under its own weight; it forms where the accumulation of snow exceeds its ablation (melting and sublimation) over many years, often centuries. Glaciers slowly deform and flow due to stresses induced by their weight, creating crevasses, seracs, and other distinguishing features. They also abrade rock and debris from their substrate to create landforms such as cirques and moraines. Glaciers form only on land and are distinct from the much thinner sea ice and lake ice that form on the surface of bodies of water.

On Earth, 99% of glacial ice is contained within vast ice sheets in the polar regions, but glaciers may be found in mountain ranges on every continent except Australia, and on a few high-latitude oceanic islands. Between 35°N and 35°S, glaciers occur only in the Himalayas, Andes, Rocky Mountains, a few high mountains in East Africa, Mexico, New Guinea and on Zard Kuh in Iran. Glaciers cover about 10 percent of Earth's land surface. Continental glaciers cover nearly 13,000,000 km² (5×10⁶ sq mi) or about 98 percent of Antarctica›s 13,200,000 km² (5.1×10⁶ sq mi), with an average thickness of 2,100 m (7,000 ft). Greenland and Patagonia also have huge expanses of continental glaciers.

Ice calving from the terminus of the Perito Moreno Glacier in western Patagonia, Argentina

The Aletsch Glacier, the largest glacier of the Alps, in Switzerland

Glacial ice is the largest reservoir of fresh water on Earth. Many glaciers from temperate, alpine and seasonal polar climates store water as ice during the colder seasons and release it later in the form of meltwater as warmer summer temperatures cause the glacier to melt, creating a water source that is especially important for plants, animals and human uses when other sources may be scant. Within high-altitude and Antarctic environments, the seasonal temperature difference is often not sufficient to release meltwater.

Because glacial mass is affected by long-term climatic changes, e.g., precipitation, mean temperature, and cloud cover, glacial mass changes are considered among the most sensitive indicators of climate change and are a major source of variations in sea level.

A large piece of compressed ice, or a glacier, appears blue, as large quantities of water appear blue. This is because water molecules absorb other colors more efficiently than blue. The other reason for the blue color of glaciers is the lack of air bubbles. Air bubbles, which give a white color to ice, are squeezed out by pressure increasing the density of the created ice.

The Quelccaya Ice Cap is the largest glaciated area in the tropics, in Peru

Etymology and Related Terms

The word *Glaceon* is a loanword from French and goes back, via Franco-Provençal, to the Vulgar Latin *glaciārium*, derived from the Late Latin *glacia*, and ultimately Latin *glaciēs*, meaning "ice". The processes and features caused by or related to glaciers are referred to as glacial. The process of glacier establishment, growth and flow is called glaciation. The corresponding area of study is called glaciology. Glaciers are important components of the global cryosphere.

Types

Classification by Size, Shape and Behavior

Mouth of the Schlatenkees Glacier near Innergschlöß, Austria

Glaciers are categorized by their morphology, thermal characteristics, and behavior. *Cirque glaciers* form on the crests and slopes of mountains. A glacier that fills a valley is called a *valley glacier*, or alternatively an *alpine glacier* or *mountain glacier*. A large body of glacial ice astride a mountain, mountain range, or volcano is termed an *ice cap* or *ice field*. Ice caps have an area less than 50,000 km² (19,000 sq mi) by definition.

Glacial bodies larger than 50,000 km² (19,000 sq mi) are called *ice sheets* or *continental glaciers*. Several kilometers deep, they obscure the underlying topography. Only nunataks protrude from their surfaces. The only extant ice sheets are the two that cover most of Antarctica and Greenland. They contain vast quantities of fresh water, enough that if both melted, global sea levels would rise by over 70 m (230 ft). Portions of an ice sheet or cap that extend into water are called ice shelves; they tend to be thin with limited slopes and reduced velocities. Narrow, fast-moving sections of an ice sheet are called *ice streams*. In Antarctica, many ice streams drain into large ice shelves. Some drain directly into the sea, often with an ice tongue, like Mertz Glacier.

The *Grotta del Gelo* is a cave of Etna volcano, the southernmost glacier in Europe

Sightseeing boat in front of a tidewater glacier, Kenai Fjords National Park, Alaska

Tidewater glaciers are glaciers that terminate in the sea, including most glaciers flowing from Greenland, Antarctica, Baffin and Ellesmere Islands in Canada, Southeast Alaska, and the Northern and Southern Patagonian Ice Fields. As the ice reaches the sea, pieces break off, or calve, forming icebergs. Most tidewater glaciers calve above sea level, which often results in a tremendous impact as the iceberg strikes the water. Tidewater glaciers undergo centuries-long cycles of advance and retreat that are much less affected by the climate change than those of other glaciers.

Classification by Thermal State

Thermally, a *temperate glacier* is at melting point throughout the year, from its surface to its base. The ice of a polar glacier is always below the freezing point from the surface to its base, although the surface snowpack may experience seasonal melting. A *sub-polar glacier* includes both temperate and polar ice, depending on depth beneath the surface and position along the length of the glacier. In a similar way, the thermal regime of a glacier is often described by the temperature at its base alone. A *cold-based glacier* is below freezing at the ice-ground interface, and is thus frozen to the underlying substrate. A *warm-based glacier* is above or at freezing at the interface, and is able to slide at this contact. This contrast is thought to a large extent to govern the ability of a glacier to effectively erode its bed, as sliding ice promotes plucking at rock from the surface below. Glaciers which are partly cold-based and partly warm-based are known as *polythermal*.

Formation

Gorner Glacier in Switzerland

A packrafter passes a wall of freshly exposed blue ice on Spencer Glacier, in Alaska. Glacial ice acts like a filter on light, and the more time light can spend traveling through ice, the bluer it becomes.

Glaciers form where the accumulation of snow and ice exceeds ablation. The area in which a glacier forms is called a cirque (corrie or cwm) – a typically armchair-shaped geological feature (such as a depression between mountains enclosed by arêtes) – which collects and compresses through gravity the snow that falls into it. This snow collects and is compacted by the weight of the snow falling above it, forming névé. Further crushing of the individual snowflakes and squeezing the air from the snow turns it into "glacial ice". This glacial ice will fill the cirque until it "overflows" through a geological weakness or vacancy, such as the gap between two mountains. When the mass of snow and ice is sufficiently thick, it begins to move due to a combination of surface slope, gravity and pressure. On steeper slopes, this can occur with as little as 15 m (50 ft) of snow-ice.

In temperate glaciers, snow repeatedly freezes and thaws, changing into granular ice called firn. Under the pressure of the layers of ice and snow above it, this granular ice fuses into denser and denser firn. Over a period of years, layers of firn undergo further compaction and become glacial ice. Glacier ice is slightly less dense than ice formed from frozen water because it contains tiny trapped air bubbles.

Glacial ice has a distinctive blue tint because it absorbs some red light due to an overtone of the infrared OH stretching mode of the water molecule. Liquid water is blue for the same reason. The blue of glacier ice is sometimes misattributed to Rayleigh scattering due to bubbles in the ice.

A glacier cave located on the Perito Moreno Glacier in Argentina

Structure

A glacier originates at a location called its glacier head and terminates at its glacier foot, snout, or terminus.

Glaciers are broken into zones based on surface snowpack and melt conditions. The ablation zone is the region where there is a net loss in glacier mass. The equilibrium line separates the ablation zone and the accumulation zone; it is the altitude where the amount of new snow gained by accumulation is equal to the amount of ice lost through ablation. The upper part of a glacier, where accumulation exceeds ablation, is called the accumulation zone. In general, the accumulation zone accounts for 60–70% of the glacier's surface area, more if the glacier calves icebergs. Ice in the accumulation zone is deep enough to exert a downward force that erodes underlying rock. After a glacier melts, it often leaves behind a bowl- or amphitheater-shaped depression that ranges in size from large basins like the Great Lakes to smaller mountain depressions known as cirques.

The accumulation zone can be subdivided based on its melt conditions.

1. The dry snow zone is a region where no melt occurs, even in the summer, and the snowpack remains dry.

2. The percolation zone is an area with some surface melt, causing meltwater to percolate into the snowpack. This zone is often marked by refrozen ice lenses, glands, and layers. The snowpack also never reaches melting point.

3. Near the equilibrium line on some glaciers, a superimposed ice zone develops. This zone is where meltwater refreezes as a cold layer in the glacier, forming a continuous mass of ice.

4. The wet snow zone is the region where all of the snow deposited since the end of the previous summer has been raised to 0 °C.

The health of a glacier is usually assessed by determining the glacier mass balance or observing terminus behavior. Healthy glaciers have large accumulation zones, more than 60% of their area snowcovered at the end of the melt season, and a terminus with vigorous flow.

Following the Little Ice Age's end around 1850, glaciers around the Earth have retreated substantially. A slight cooling led to the advance of many alpine glaciers between 1950 and 1985, but since 1985 glacier retreat and mass loss has become larger and increasingly ubiquitous.

Motion

Shear or herring-bone crevasses on Emmons Glacier (Mount Rainier); such crevasses often form near the edge of a glacier where interactions with underlying or marginal rock impede flow. In this case, the impediment appears to be some distance from the near margin of the glacier.

Glaciers move, or flow, downhill due to gravity and the internal deformation of ice. Ice behaves like a brittle solid until its thickness exceeds about 50 m (160 ft). The pressure on ice deeper than 50 m causes plastic flow. At the molecular level, ice consists of stacked layers of molecules with relatively weak bonds between layers. When the stress on the layer above exceeds the inter-layer binding strength, it moves faster than the layer below.

Glaciers also move through basal sliding. In this process, a glacier slides over the terrain on which it sits, lubricated by the presence of liquid water. The water is created from ice that melts under high pressure from frictional heating. Basal sliding is dominant in temperate, or warm-based glaciers.

Perito Moreno glacier

Fracture Zone and Cracks

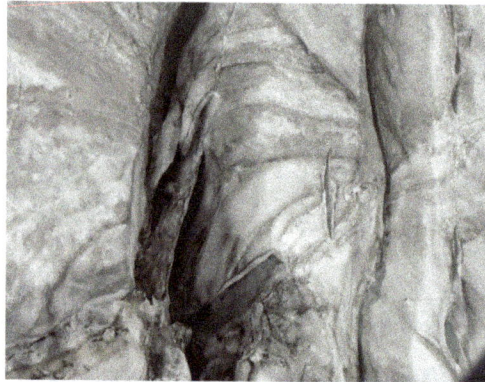
Ice cracks in the Titlis Glacier

The top 50 m (160 ft) of a glacier are rigid because they are under low pressure. This upper section is known as the *fracture zone* and moves mostly as a single unit over the plastically flowing lower section. When a glacier moves through irregular terrain, cracks called crevasses develop in the fracture zone. Crevasses form due to differences in glacier velocity. If two rigid sections of a glacier move at different speeds and directions, shear forces cause them to break apart, opening a crevasse. Crevasses are seldom more than 46 m (150 ft) deep but in some cases can be 300 m (1,000 ft) or even deeper. Beneath this point, the plasticity of the ice is too great for cracks to form. Intersecting crevasses can create isolated peaks in the ice, called seracs.

Crevasses can form in several different ways. Transverse crevasses are transverse to flow and form where steeper slopes cause a glacier to accelerate. Longitudinal crevasses form semi-parallel to flow where a glacier expands laterally. Marginal crevasses form from the edge of the glacier, due to the reduction in speed caused by friction of the valley walls. Marginal crevasses are usually largely transverse to flow. Moving glacier ice can sometimes separate from stagnant ice above, forming a bergschrund. Bergschrunds resemble crevasses but are singular features at a glacier's margins.

Crevasses make travel over glaciers hazardous, especially when they are hidden by fragile snow bridges.

Crossing a crevasse on the Easton Glacier, Mount Baker, in the North Cascades, United States

Below the equilibrium line, glacial meltwater is concentrated in stream channels. Meltwater can pool in proglacial lakes on top of a glacier or descend into the depths of a glacier via moulins. Streams within or beneath a glacier flow in englacial or sub-glacial tunnels. These tunnels sometimes reemerge at the glacier's surface.

Speed

The speed of glacial displacement is partly determined by friction. Friction makes the ice at the bottom of the glacier move more slowly than ice at the top. In alpine glaciers, friction is also generated at the valley's side walls, which slows the edges relative to the center.

Mean speeds vary greatly, but is typically around 1 m (3 ft) per day. There may be no motion in stagnant areas; for example, in parts of Alaska, trees can establish themselves on surface sediment deposits. In other cases, glaciers can move as fast as 20–30 m (70–100 ft) per day, such as in Greenland's Jakobshavn Isbræ (Greenlandic: *Sermeq Kujalleq*). Velocity increases with increasing slope, increasing thickness, increasing snowfall, increasing longitudinal confinement, increasing basal temperature, increasing meltwater production and reduced bed hardness.

A few glaciers have periods of very rapid advancement called surges. These glaciers exhibit normal movement until suddenly they accelerate, then return to their previous state. During these surges, the glacier may reach velocities far greater than normal speed. These surges may be caused by failure of the underlying bedrock, the pooling of meltwater at the base of the glacier — perhaps delivered from a supraglacial lake — or the simple accumulation of mass beyond a critical «tipping point». Temporary rates up to 90 m (300 ft) per day have occurred when increased temperature or overlying pressure caused bottom ice to melt and water to accumulate beneath a glacier.

In glaciated areas where the glacier moves faster than one km per year, glacial earthquakes occur. These are large scale earthquakes that have seismic magnitudes as high as 6.1. The number of glacial earthquakes in Greenland peaks every year in July, August and September and is increasing over time. In a study using data from January 1993 through October 2005, more events were detected every year since 2002, and twice as many events were recorded in 2005 as there were in any other year. This increase in the numbers of glacial earthquakes in Greenland may be a response to global warming.

Ogives

Ogives are alternating wave crests and valleys that appear as dark and light bands of ice on glacier surfaces. They are linked to seasonal motion of glaciers; the width of one dark and one light band generally equals the annual movement of the glacier. Ogives are formed when ice from an icefall is severely broken up, increasing ablation surface area during summer. This creates a swale and space for snow accumulation in the winter, which in turn creates a ridge. Sometimes ogives consist only of undulations or color bands and are described as wave ogives or band ogives.

Geography

Black ice glacier near Aconcagua, Argentina

Glaciers are present on every continent and approximately fifty countries, excluding those (Australia, South Africa) that have glaciers only on distant subantarctic island territories. Extensive glaciers are found in Antarctica, Chile, Canada, Alaska, Greenland and Iceland. Mountain glaciers are widespread, especially in the Andes, the Himalayas, the Rocky Mountains, the Caucasus, Scandinavian mountains and the Alps. Mainland

Australia currently contains no glaciers, although a small glacier on Mount Kosciuszko was present in the last glacial period. In New Guinea, small, rapidly diminishing, glaciers are located on its highest summit massif of Puncak Jaya. Africa has glaciers on Mount Kilimanjaro in Tanzania, on Mount Kenya and in the Rwenzori Mountains. Oceanic islands with glaciers include Iceland, several of the islands off the coast of Norway including Svalbard and Jan Mayen to the far North, New Zealand and the subantarctic islands of Marion, Heard, Grande Terre (Kerguelen) and Bouvet. During glacial periods of the Quaternary, Taiwan, Hawaii on Mauna Kea and Tenerife also had large alpine glaciers, while the Faroe and Crozet Islands were completely glaciated.

The permanent snow cover necessary for glacier formation is affected by factors such as the degree of slope on the land, amount of snowfall and the winds. Glaciers can be found in all latitudes except from 20° to 27° north and south of the equator where the presence of the descending limb of the Hadley circulation lowers precipitation so much that with high insolation snow lines reach above 6,500 m (21,330 ft). Between 19°N and 19°S, however, precipitation is higher and the mountains above 5,000 m (16,400 ft) usually have permanent snow.

Even at high latitudes, glacier formation is not inevitable. Areas of the Arctic, such as Banks Island, and the McMurdo Dry Valleys in Antarctica are considered polar deserts where glaciers cannot form because they receive little snowfall despite the bitter cold. Cold air, unlike warm air, is unable to transport much water vapor. Even during glacial periods of the Quaternary, Manchuria, lowland Siberia, and central and northern Alaska, though extraordinarily cold, had such light snowfall that glaciers could not form.

In addition to the dry, unglaciated polar regions, some mountains and volcanoes in Bolivia, Chile and Argentina are high (4,500 to 6,900 m or 14,800 to 22,600 ft) and cold, but the relative lack of precipitation prevents snow from accumulating into glaciers. This is because these peaks are located near or in the hyperarid Atacama Desert.

Glacial Geology

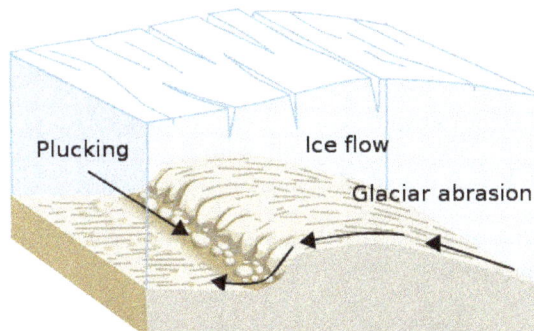

Diagram of glacial plucking and abrasion

As glaciers flow over bedrock, they soften and lift blocks of rock into the ice. This process, called plucking, is caused by subglacial water that penetrates fractures in the bedrock and subsequently freezes and expands. This expansion causes the ice to act as a lever that loosens the rock by lifting it. Thus, sediments of all sizes become part of the glacier's load. If a retreating glacier gains enough debris, it may become a rock glacier, like the Timpanogos Glacier in Utah.

Glacially pl.ucked granitic bedrock near Mariehamn, Åland Islands

Glaciers erode terrain through two principal processes: abrasion and plucking.

Abrasion occurs when the ice and its load of rock fragments slide over bedrock and function as sandpaper, smoothing and polishing the bedrock below. The pulverized rock this process produces is called rock flour and is made up of rock grains between 0.002 and 0.00625 mm in size. Abrasion leads to steeper valley walls and mountain slopes in alpine settings, which can cause avalanches and rock slides, which add even more material to the glacier.

Glacial abrasion is commonly characterized by glacial striations. Glaciers produce these when they contain large boulders that carve long scratches in the bedrock. By mapping the direction of the striations, researchers can determine the direction of the glacier's movement. Similar to striations are chatter marks, lines of crescent-shape depressions in the rock underlying a glacier. They are formed by abrasion when boulders in the glacier are repeatedly caught and released as they are dragged along the bedrock.

The rate of glacier erosion varies. Six factors control erosion rate:

- Velocity of glacial movement

- Thickness of the ice

- Shape, abundance and hardness of rock fragments contained in the ice at the bottom of the glacier

- Relative ease of erosion of the surface under the glacier

- Thermal conditions at the glacier base

- Permeability and water pressure at the glacier base

When the bedrock has frequent fractures on the surface, glacial erosion rates tend to increase as plucking is the main erosive force on the surface; when the bedrock has wide gaps between sporadic fractures, however, abrasion tends to be the dominant erosive form and glacial erosion rates become slow.

Glaciers in lower latitudes tend to be much more erosive than glaciers in higher latitudes, because they have more meltwater reaching the glacial base and facilitate sediment production and transport under the same moving speed and amount of ice.

Material that becomes incorporated in a glacier is typically carried as far as the zone of ablation before being deposited. Glacial deposits are of two distinct types:

- *Glacial till*: material directly deposited from glacial ice. Till includes a mixture of undifferentiated material ranging from clay size to boulders, the usual composition of a moraine.

- *Fluvial and outwash sediments*: sediments deposited by water. These deposits are stratified by size.

Larger pieces of rock that are encrusted in till or deposited on the surface are called "glacial erratics". They range in size from pebbles to boulders, but as they are often moved great distances, they may be drastically different from the material upon which they are found. Patterns of glacial erratics hint at past glacial motions.

Moraines

Glacial moraines above Lake Louise, Alberta, Canada

Glacial moraines are formed by the deposition of material from a glacier and are exposed after the glacier has retreated. They usually appear as linear mounds of till, a non-sorted mixture of rock, gravel and boulders within a matrix of a fine powdery material. Terminal or end moraines are formed at the foot or terminal end of a glacier. Lateral moraines are formed on the sides of the glacier. Medial moraines are formed when two different glaciers merge and the lateral moraines of each coalesce to form a moraine in the middle of the combined glacier. Less apparent are ground moraines, also called *glacial drift*, which often blankets the surface underneath the glacier downslope from the equilibrium line.

The term *moraine* is of French origin. It was coined by peasants to describe alluvial embankments and rims found near the margins of glaciers in the French Alps. In modern geology, the term is used more broadly, and is applied to a series of formations, all of which are composed of till. Moraines can also create moraine dammed lakes.

Drumlins

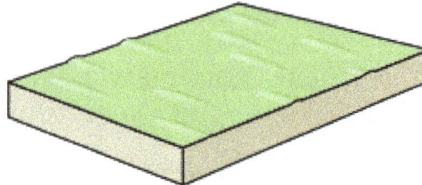

A drumlin field forms after a glacier has modified the landscape. The teardrop-shaped formations denote the direction of the ice flow.

Drumlins are asymmetrical, canoe shaped hills made mainly of till. Their heights vary from 15 to 50 meters and they can reach a kilometer in length. The steepest side of the hill faces the direction from which the ice advanced (*stoss*), while a longer slope is left in the ice's direction of movement (*lee*).

Drumlins are found in groups called *drumlin fields* or *drumlin camps*. One of these fields is found east of Rochester, New York; it is estimated to contain about 10,000 drumlins.

Although the process that forms drumlins is not fully understood, their shape implies that they are products of the plastic deformation zone of ancient glaciers. It is believed that many drumlins were formed when glaciers advanced over and altered the deposits of earlier glaciers.

Glacial Valleys, Cirques, Arêtes, and Pyramidal Peaks

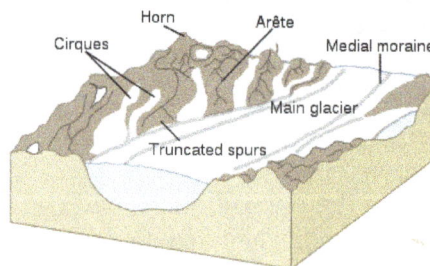

Features of a glacial landscape

Before glaciation, mountain valleys have a characteristic "V" shape, produced by eroding water. During glaciation, these valleys are often widened, deepened and smoothed to form a "U"-shaped glacial valley. The erosion that creates glacial valleys truncates any spurs of rock or earth that may have earlier extended across the valley, creating broadly triangular-shaped cliffs called truncated spurs. Within glacial valleys, depressions created by plucking and abrasion can be filled by lakes, called paternoster lakes. If a glacial valley runs into a large body of water, it forms a fjord.

Typically glaciers deepen their valleys more than their smaller tributaries. Therefore, when glaciers recede, the valleys of the tributary glaciers remain above the main glacier's depression and are called hanging valleys.

At the start of a classic valley glacier is a bowl-shaped cirque, which has escarped walls on three sides but is open on the side that descends into the valley. Cirques are where ice begins to accumulate in a glacier. Two glacial cirques may form back to back and erode their backwalls until only a narrow ridge, called an arête is left. This structure may result in a mountain pass. If multiple cirques encircle a single mountain, they create pointed pyramidal peaks; particularly steep examples are called horns.

Roches Moutonnées

Passage of glacial ice over an area of bedrock may cause the rock to be sculpted into a knoll called a *roche moutonnée,* or "sheepback" rock. Roches moutonnées may be elongated, rounded and asymmetrical in shape. They range in length from less than a meter to several hundred meters long. Roches moutonnées have a gentle slope on their up-glacier sides and a steep to vertical face on their down-glacier sides. The glacier abrades the smooth slope on the upstream side as it flows along, but tears rock fragments loose and carries them away from the downstream side via plucking.

Alluvial Stratification

As the water that rises from the ablation zone moves away from the glacier, it carries fine eroded sediments with it. As the speed of the water decreases, so does its capacity to carry objects in suspension. The water thus gradually deposits the sediment as it runs, creating an alluvial plain. When this phenomenon occurs in a valley, it is called a *valley train*. When the deposition is in an estuary, the sediments are known as bay mud.

Outwash plains and valley trains are usually accompanied by basins known as "kettles". These are small lakes formed when large ice blocks that are trapped in alluvium melt and produce water-filled depressions. Kettle diameters range from 5 m to 13 km, with depths of up to 45 meters. Most are circular in shape because the blocks of ice that formed them were rounded as they melted.

Glacial Deposits

When a glacier's size shrinks below a critical point, its flow stops and it becomes stationary. Meanwhile, meltwater within and beneath the ice leaves stratified alluvial deposits. These deposits, in the forms of columns, terraces and clusters, remain after the glacier melts and are known as "glacial deposits".

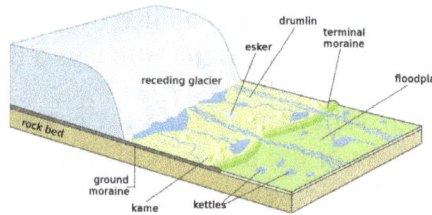

Landscape produced by a receding glacier

Glacial deposits that take the shape of hills or mounds are called *kames*. Some kames form when meltwater deposits sediments through openings in the interior of the ice. Others are produced by fans or deltas created by meltwater. When the glacial ice occupies a valley, it can form terraces or kames along the sides of the valley.

Long, sinuous glacial deposits are called *eskers*. Eskers are composed of sand and gravel that was deposited by meltwater streams that flowed through ice tunnels within or beneath a glacier. They remain after the ice melts, with heights exceeding 100 meters and lengths of as long as 100 km.

Loess Deposits

Very fine glacial sediments or rock flour is often picked up by wind blowing over the bare surface and may be deposited great distances from the original fluvial deposition site. These eolian loess deposits may be very deep, even hundreds of meters, as in areas of China and the Midwestern United States of America. Katabatic winds can be important in this process.

Isostatic Rebound

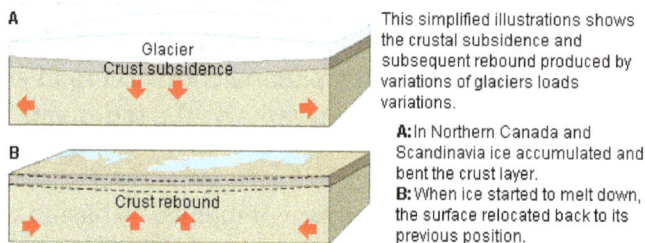

This simplified illustrations shows the crustal subsidence and subsequent rebound produced by variations of glaciers loads variations.

A: In Northern Canada and Scandinavia ice accumulated and bent the crust layer.
B: When ice started to melt down, the surface relocated back to its previous position.

Isostatic pressure by a glacier on the Earth's crust

Large masses, such as ice sheets or glaciers, can depress the crust of the Earth into the mantle. The depression usually totals a third of the ice sheet or glacier's thickness. After the ice sheet or glacier melts, the mantle begins to flow back to its original position, pushing the crust back up. This post-glacial rebound, which proceeds very slowly after the melting of the ice sheet or glacier, is currently occurring in measurable amounts in Scandinavia and the Great Lakes region of North America.

A geomorphological feature created by the same process on a smaller scale is known as *dilation-faulting*. It occurs where previously compressed rock is allowed to return to its original shape more rapidly than can be maintained without faulting. This leads to an effect similar to what would be seen if the rock were hit by a large hammer. Dilation faulting can be observed in recently de-glaciated parts of Iceland and Cumbria.

On Mars

Northern polar ice cap on Mars

The polar ice caps of Mars show geologic evidence of glacial deposits. The south polar cap is especially comparable to glaciers on Earth. Topographical features and computer models indicate the existence of more glaciers in Mars' past.

At mid-latitudes, between 35° and 65° north or south, Martian glaciers are affected by the thin Martian atmosphere. Because of the low atmospheric pressure, ablation near the surface is solely due to sublimation, not melting. As on Earth, many glaciers are covered with a layer of rocks which insulates the ice. A radar instrument on board the Mars Reconnaissance Orbiter found ice under a thin layer of rocks in formations called lobate debris aprons (LDAs).

The pictures below illustrate how landscape features on Mars closely resemble those on the Earth.

Romer Lake's Elephant Foot Glacier in the Earth's Arctic, as seen by Landsat 8. This picture shows several glaciers that have the same shape as many features on Mars that are believed to also be glaciers. The next three images from Mars show shapes similar to the Elephant Foot Glacier.

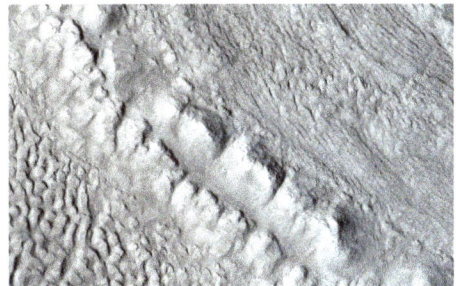

Enlargement of area in rectangle of the previous image. On Earth the ridge would be called the terminal moraine of an alpine glacier. Picture taken with HiRISE under the HiWish program. Image from Ismenius Lacus quadrangle.

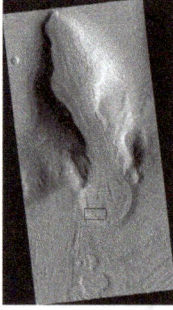

Glacier as seen by HiRISE under the HiWish program. Area in rectangle is enlarged in the next photo. Zone of accumulation of snow at the top. Glacier is moving down valley, then spreading out on plain. Evidence for flow comes from the many lines on surface. Location is in Protonilus Mensae in Ismenius Lacus quadrangle

Mesa in Ismenius Lacus quadrangle, as seen by CTX. Mesa has several glaciers eroding it. One of the glaciers is seen in greater detail in the next two images from HiRISE. Image from Ismenius Lacus quadrangle.

Iceberg

Iceberg in the Arctic with its underside visible.

An iceberg or ice mountain is a large piece of freshwater ice that has broken off a glacier or an ice shelf and is floating freely in open water. It may subsequently become frozen into pack ice (one form of sea ice). As it drifts into shallower waters, it may come into contact with the seabed, a process referred to as seabed gouging by ice. Almost 91% of an iceberg is below the surface of the water.

Etymology

The word "iceberg" is a partial loan translation from Dutch *ijsberg*, literally meaning *ice mountain*, cognate to Danish *isbjerg*, German *Eisberg*, Low Saxon *Iesbarg* and Swedish *isberg*.

Overview

Because the density of pure ice is about 920 kg/m³, and that of seawater about 1025 kg/m³, typically only one-tenth of the volume of an iceberg is above water (due to Archimedes's Principle). The shape of the underwater portion can be difficult to judge by looking at the portion above the surface. This has led to the expression "tip of the iceberg", for a problem or difficulty that is only a small manifestation of a larger problem.

Grotto in an iceberg, photographed during the British Antarctic Expedition of 1911–1913, 5 Jan 1911. Photographer: Herbert Ponting, Alexander Turnbull Library

Icebergs generally range from 1 to 75 metres (3.3 to 246.1 ft) above sea level and weigh 100,000 to 200,000 metric tons (110,000 to 220,000 short tons). The largest known iceberg in the North Atlantic was 168 metres (551 ft) above sea level, reported by the USCG icebreaker *East Wind* in 1958, making it the height of a 55-story building. These icebergs originate from the glaciers of western Greenland and may have an interior temperature of –15 to –20 °C (5 to –4 °F).

Icebergs are usually confined by winds and currents to move close to the coast. The largest icebergs recorded have been calved, or broken off, from the Ross Ice Shelf of Antarctica. Iceberg B-15, photographed by satellite in 2000, measured 295 by 37 kilometres (183 by 23 mi), with a surface area of 11,000 square kilometres (4,200 sq mi). The largest iceberg on record was an Antarctic tabular iceberg of over 31,000 square kilometres (12,000 sq mi) [335 by 97 kilometres (208 by 60 mi)] sighted 150 miles (240 km) west of Scott Island, in the South Pacific Ocean, by the USS *Glacier* on November 12, 1956. This iceberg was larger than Belgium.

When a piece of iceberg ice melts, it makes a fizzing sound called "Bergie Seltzer". This sound is made when the water-ice interface reaches compressed air bubbles trapped in the ice. As this happens, each bubble bursts, making a 'popping' sound. The bubbles contain air trapped in snow layers very early in the history of the ice, that eventually got buried to a given depth (up to several kilometers) and pressurized as it transformed into firn then to glacial ice.

Recent large Icebergs

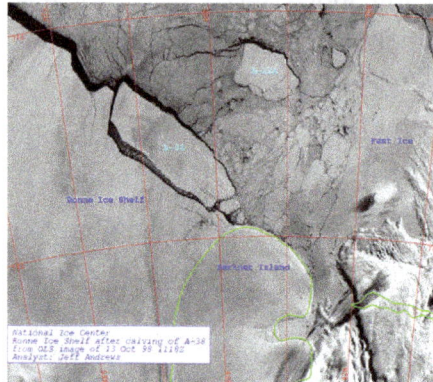

The calving of Iceberg A-38 off Ronne Ice Shelf

- Iceberg B-15 11,000 km² (4,200 sq mi), 2000

- Iceberg A-38, about 6,900 km² (2,700 sq mi), 1998

- Iceberg B-15A, 3,100 km² (1,200 sq mi), broke off 2003

- Iceberg C-19, 5,500 km² (2,100 sq mi), 2002

- Iceberg B-9, 5,390 km² (2,080 sq mi), 1987

- Iceberg B-31, 615 km² (237 sq mi), 2014

- Iceberg D-16, 310 km² (120 sq mi), 2006

- Ice sheet, 260 km² (100 sq mi), broken off of Petermann Glacier in northern Greenland on Aug 5, 2010, considered to be the largest Arctic iceberg since 1962. About a month later, this iceberg split into two pieces upon crashing into Joe Island in the Nares Strait next to Greenland. In June 2011, large fragments of the Petermann Ice Islands were observed off the Labrador coast.

- Iceberg B-17B 140 km² (54 sq mi), 1999, shipping alert issued December 2009.

Shape

Tabular iceberg, near Brown Bluff in the Antarctic Sound off Tabarin Peninsula

Non-tabular iceberg off Elephant Island in the Southern Ocean

In addition to size classification, icebergs can be classified on the basis of their shape. The two basic types of iceberg forms are *tabular* and *non-tabular*. Tabular icebergs have steep sides and a flat top, much like a plateau, with a length-to-height ratio of more than 5:1. This type of iceberg, also known as an *ice island*, can be quite large, as in the case of Pobeda Ice Island. Antarctic icebergs formed by breaking off from an ice shelf, such as the Ross Ice Shelf or Filchner-Ronne Ice Shelf, are typically tabular. The largest icebergs in the world are formed this way.

Non-tabular icebergs have different shapes and include:

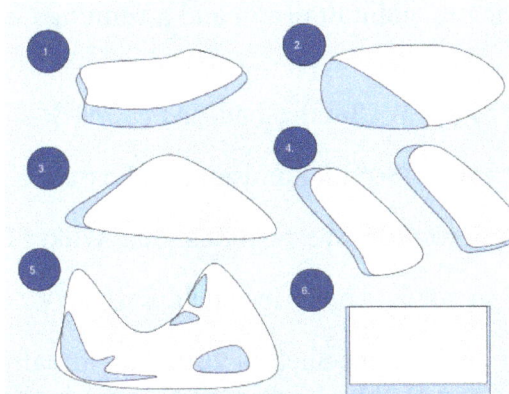
Different shapes of icebergs. 1: Tabular; 2: Wedge; 3: Dome; 4: Drydock; 5: Pinnacled; 6: Blocky.

- *Dome*: An iceberg with a rounded top.

- *Pinnacle*: An iceberg with one or more spires.

- *Wedge*: An iceberg with a steep edge on one side and a slope on the opposite side.

- *Dry-Dock*: An iceberg that has eroded to form a slot or channel.

- *Blocky*: An iceberg with steep, vertical sides and a flat top. It differs from tabular icebergs in that its shape is more like a block than a flat sheet.

Monitoring

Icebergs are monitored worldwide by the U.S. National Ice Center (NIC), established in 1995, which produces analyses and forecasts of Arctic, Antarctic, Great Lakes and Chesapeake Bay ice conditions. More than 95% of the data used in its sea ice analyses are derived from the remote sensors on polar-orbiting satellites that survey these remote regions of the Earth.

Iceberg A22A in the South Atlantic Ocean

The NIC is the only organization that names and tracks all Antarctic Icebergs. It assigns each iceberg larger than 10 nautical miles (19 km) along at least one axis a name composed of a letter indicating its point of origin and a running number. The letters used are as follows:

   A – longitude 0° to 90° W (Bellingshausen Sea, Weddell Sea)

   B – longitude 90° W to 180° (Amundsen Sea, Eastern Ross Sea)

   C – longitude 90° E to 180° (Western Ross Sea, Wilkes Land)

   D – longitude 0° to 90° E (Amery Ice Shelf, Eastern Weddell Sea)

Iceberg B15 calved from the Ross Ice Shelf in 2000 and initially had an area of 11,000 square kilometres (4,200 sq mi). It broke apart in November 2002. The largest remaining piece of it, Iceberg B-15A, with an area of 3,000 square kilometres (1,200 sq mi), was still the largest iceberg on Earth until it ran aground and split into several pieces October 27, 2005, an event that was observed by seismographs both on the iceberg and across Antarctica. It has been hypothesized that this breakup may also have been abetted by ocean swell generated by an Alaskan storm 6 days earlier and 13,500 kilometres (8,400 mi) away.

History

In the 20th century, several scientific bodies were established to study and monitor the icebergs. The International Ice Patrol, formed in 1914 in response to the April 1912 sinking of the *Titanic*, which killed 1,517 of its 2,223 passengers, monitors iceberg dangers near the Grand Banks of Newfoundland and provides the "limits of all known ice" in that vicinity to the maritime community.

The iceberg suspected of sinking the RMS Titanic; a smudge of red paint much like the Titanic's red hull stripe was seen near its base at the waterline.

Technology History

Before the early 1910s there was no system in place to track icebergs to guard ships against collisions, most likely because they weren't considered a serious threat back then, ships have managed to survive even direct crashes. In 1907 *SS Kronprinz Wilhelm*, a German liner, had rammed an iceberg and suffered a crushed bow, but was still able to complete her voyage. The April 1912 sinking of the *Titanic* however changed all that, and created the demand for a system to observe icebergs. For the remainder of the ice season of that year, the United States Navy patrolled the waters and monitored ice flow. In November 1913, the International Conference on the Safety of Life at Sea met in London to devise a more permanent system of observing icebergs. Within three months the participating maritime nations had formed the International Ice Patrol (IIP). The goal of the IIP was to collect data on meteorology and oceanography to measure currents, ice-flow, ocean temperature, and salinity levels. They published their first records in 1921, which allowed for a year-by-year comparison of iceberg movement.

New technologies monitor icebergs. Aerial surveillance of the seas in the early 1930s allowed for the development of charter systems that could accurately detail the ocean currents and iceberg locations. In 1945, experiments tested the effectiveness of radar in detecting icebergs. A decade later, oceanographic monitoring outposts were established for the purpose of collecting data; these outposts continue to serve in environmental study. A computer was first installed on a ship for the purpose of oceanographic monitoring in 1964, which allowed for a faster evaluation of data. By the 1970s, icebreaking ships were equipped with automatic transmissions of satellite photographs of ice in Antarctica. Systems for optical satellites had been developed but were still limited by weather conditions. In the 1980s, drifting buoys were used in Antarctic waters for oceanographic and climate research. They are equipped with sensors that measure ocean temperature and currents.

An iceberg being pushed by three U.S. Navy ships in McMurdo Sound, Antarctica

Side looking airborne radar (SLAR) made it possible to acquire images regardless of weather conditions. On November 4, 1995, Canada launched RADARSAT-1. Developed by the Canadian Space Agency, it provides images of Earth for scientific and commercial purposes. This system was the first to use synthetic aperture radar (SAR), which sends microwave energy to the ocean surface and records the reflections to track icebergs. The European Space Agency launched ENVISAT (an observation satellite that orbits the Earth's poles) on March 1, 2002. ENVISAT employs advanced synthetic aperture radar (ASAR) technology, which can detect changes in surface height accurately. The Canadian Space Agency launched RADARSAT-2 in December 2007, which uses SAR and multi-polarization modes and follows the same orbit path as RADARSAT-1.

Bog

Mer Bleue Bog, a typical peat bog, in Ontario, Canada.

A bog is a wetland that accumulates peat, a deposit of dead plant material—often mosses, and in a majority of cases, sphagnum moss. It is one of the four main types of wetlands. Other names for bogs include mire, quagmire, and muskeg; alkaline mires are called fens. They are frequently covered in ericaceous shrubs rooted in the sphagnum moss and peat. The gradual accumulation of decayed plant material in a bog functions as a carbon sink.

Bogs occur where the water at the ground surface is acidic and low in nutrients. In some cases, the water is derived entirely from precipitation, in which case they are termed ombrotrophic (rain-fed). Water flowing out of bogs has a characteristic brown colour, which comes from dissolved peat tannins. In general, the low fertility and cool climate results in relatively slow plant growth, but decay is even slower owing to the saturated soil. Hence peat accumulates. Large areas of landscape can be covered many metres deep in peat.

Precipitation accumulates in many bogs, forming bog pools, such as Koitjärve bog in Estonia

Bogs have distinctive assemblages of animal, fungal and plant species, and are of high importance for biodiversity, particularly in landscapes that are otherwise settled and farmed.

Distribution and Extent

Carnivorous plants, such as this *Sarracenia purpurea* pitcher plant of the eastern seaboard of North America, are often found in bogs. Capturing insects provides nitrogen and phosphorus, which are usually scarce in such conditions.

Bogs are widely distributed in cold, temperate climes, mostly in boreal ecosystems in the Northern Hemisphere. The world's largest wetland is the peat bogs of the Western Siberian Lowlands in Russia, which cover more than a million square kilometres. Large peat bogs also occur in North America, particularly the Hudson Bay Lowland and the Mackenzie River Basin. They are less common in the Southern Hemisphere, with the largest being the Magellanic moorland, comprising some 44,000 square kilometres. Sphagnum bogs were widespread in northern Europe but have often been cleared and drained for agriculture.

A 2014 expedition leaving from Itanga village, Republic of the Congo discovered a peat bog "as big as England" which stretches into neighboring Democratic Republic of Congo.

Habitats

An expanse of wet Sphagnum bog in Frontenac National Park, Quebec, Canada. Spruce trees can be seen on a forested ridge in the background.

There are many highly specialised animals, fungi and plants associated with bog habitat. Most are capable of tolerating the combination of low nutrient levels and waterlogging. Sphagnum moss is generally abundant, along with ericaceous shrubs. The shrubs are often evergreen, which is understood to assist in conservation of nutrients. In drier locations, evergreen trees can occur, in which case the bog blends into the surrounding expanses of boreal evergreen forest. Sedges are one of the more common herbaceous species. Carnivorous plants such as sundews (*Drosera*) and pitcher plants (for example *Sarracenia purpurea*) have adapted to the low-nutrient conditions by using invertebrates as a nutrient source. Orchids have adapted to these conditions through the use of mycorrhizal fungi to extract nutrients. Some shrubs such as *Myrica gale* (bog myrtle) have root nodules in which nitrogen fixation occurs, thereby providing another supplemental source of nitrogen.

Many species of evergreen shrub are found in bogs, such as Labrador tea.

Bogs are recognized as a significant/specific habitat type by a number of governmental and conservation agencies. They can provide habitat for mammals, such as caribou, moose, and beavers, as well as for species of nesting shorebirds, such as Siberian cranes and yellowlegs. The United Kingdom in its Biodiversity Action Plan establishes bog habitats as a priority for conservation. Russia has a large reserve system in the West Siberian Lowland. The highest protected status occurs in Zapovedniks (IUCN category IV); Gydansky and Yugansky are two prominent examples. Bogs even have distinctive insects; English bogs give a home to a yellow fly called the hairy canary fly (*Phaonia jaroschewskii*), and bogs in North America are habitat for a butterfly called the bog copper (*Lycaena epixanthe*).

Types

Bog habitats may develop in various situations, depending on the climate and topography.

By Location and Water Source

One way of classifying them is based upon their location in the landscape, and their source of water.

Valley Bog

Aerial image of Carbajal Valley peat bogs, Tierra del Fuego Province, Argentina.

These develop in gently sloping valleys or hollows. A layer of peat fills the deepest part of the valley, and a stream may run through the surface of the bog. Valley bogs may develop in relatively dry and warm climates, but because they rely on ground or surface water, they only occur on acidic substrates.

Raised Bog

These develop from a lake or flat marshy area, over either non-acidic or acidic substrates. Over centuries there is a progression from open lake, to marsh, then fen (or on acidic substrates, valley bog) and carr, as silt or peat fills the lake. Eventually peat builds up to a level where the land surface is too flat for ground or surface water to reach the centre of the wetland. This part therefore becomes wholly rain-fed (ombrotrophic), and the resulting acidic conditions allow the development of bog (even if the substrate is non-acidic). The bog continues to form peat, and over time a shallow dome of bog peat develops: a raised bog. The dome is typically a few metres high in the centre, and is often surrounded by strips of fen or other wetland vegetation at the edges or along streamsides, where ground water can percolate into the wetland.

Viru Bog in Lahemaa National Park, Estonia, which is rich in raised bogs.

The various types of raised bog may be divided into:

- Coastal bog

- Plateau bog

- Upland bog

- Kermi bog

- String bog

- Palsa bog

- Polygonal bog

Blanket Bog

Sphagnum moss and sedges can produce floating bog mats along the shores of small lakes. This bog in Duck Lake, Oregon also supports a carnivorous plant, sundew.

Blanket bog in Connemara, Ireland

In cool climates with consistently high rainfall (on more than c. 235 days a year), the ground surface may remain waterlogged for much of the time, providing conditions for the development of bog vegetation. In these circumstances bog develops as a layer "blanketing" much of the land, including hilltops and slopes. Although a blanket bog is more common on acidic substrates, under some conditions it may also develop on neutral or even alkaline ones, if abundant acidic rainwater predominates over the ground water. A blanket bog cannot occur in drier or warmer climates, because under those conditions hilltops and sloping ground dry out too often for peat to form – in intermediate climates a blanket bog may be limited to areas which are shaded from direct sunshine. In periglacial climates a patterned form of blanket bog may occur, known as a string bog. In Europe, these mostly very thin peat layers without significant surface structures are distributed over the hills and valleys of Ireland, Scotland, England and Norway. In North America, blanket bogs occur predominantly in Canada east of Hudson Bay. These bogs are often still under the influence of mineral soil water (groundwater). Blanket bogs do not occur north of the 65th latitude in the northern hemisphere.

Quaking Bog

A *quaking bog* or *schwingmoor* is a form of bog occurring in wetter parts of valley bogs and raised bogs, and sometimes around the edges of acidic lakes. The bog vegetation, mostly sphagnum moss anchored by sedges (such as *Carex lasiocarpa*), forms a floating mat approximately half a metre thick, on the surface of the water or on top of very wet peat. White spruces are also common in this bog regime. Walking on the surface causes it to move – larger movements may cause visible ripples on the surface, or they may even make trees sway. In the absence of disturbance from waves, the bog mat may eventually cover entire bays, or even entire small lakes.

Cataract Bog

A cataract bog is a rare ecological community formed where a permanent stream flows over a granite outcropping. The sheeting of water keeps the edges of the rock wet without eroding the soil, but in this precarious location no tree or large shrub can maintain a roothold. The result is a narrow, permanently wet habitat.

By Nutrient Content

Bogs may also be classified by the nutrient content of the peat.

Eutrophic Bog

A eutrophic bog, also called a minerotrophic bog, is one that lies on top of fen-peat. As a result its water is rich in nutrients. They are found in temperate regions. Fens are eutrophic lowland bogs.

Mesotrophic Bog

A mesotrophic bog, also called a transitional peat bog, contains a moderate quantity of nutrients.

Oligotrophic Bog

Oligotrophic bogs occur where the groundwater is poor in nutrients e.g. in wetlands with nutrient-poor soils. They occur in several variants: raised bogs, soligenic bogs and blanket bog.

Uses

Industrial Uses

Sitniki peat bog in Russia recultivated after industrial use.

After drying, peat is used as a fuel, and it has been used that way for centuries. More than 20% of home heat in Ireland comes from peat, and it is also used for fuel in Finland, Scotland, Germany, and Russia. Russia is the leading exporter of peat for fuel, at more than 90 million metric tons per year. Ireland's *Bord na Móna* ("peat board") was one of the first companies to mechanically harvest peat, which is being phased out.

The other major use of dried peat is as a soil amendment (sold as *moss peat* or *sphagnum peat*) to increase the soil's capacity to retain moisture and enrich the soil. It is also used as a mulch. Some distilleries, notably in the Islay whisky-producing region, use the smoke from peat fires to dry the barley used in making Scotch whisky.

Once the peat has been extracted, it can be difficult to restore the wetland, since peat accumulation is a slow process. More than 90% of the bogs in England have been damaged or destroyed. In 2011 plans for elimination of peat in gardening products were announced by the U.K. government.

Other uses

The peat in bogs is an important place for the storage of carbon. If the peat decayed, carbon dioxide would be released to the atmosphere, contributing to global warming. Undisturbed, bogs function as a carbon sink. As one example, the peatlands of the former Soviet Union were calculated to be removing 52 Tg of carbon per year from the atmosphere.

Peat bogs are also important in storing fresh water, particularly in the headwaters of large rivers. Even the enormous Yangtze River arises in the Ruoergai peatland near its headwaters in Tibet.

Blueberries, cranberries, cloudberries, huckleberries, and lingonberries are harvested from the wild in bogs. Bog oak, wood that has been partially preserved by bogs, has been used in the manufacture of furniture.

Sphagnum bogs are also used for outdoor recreation, with activities including ecotourism and hunting. For example, many popular canoe routes in northern Canada include areas of peatland. Some other activities, such as all-terrain vehicle use, are especially damaging to bogs.

Archaeology

The anaerobic environment and presence of tannic acids within bogs can result in the remarkable preservation of organic material. Finds of such material have been made in Denmark, Germany, Ireland, Russia, and the United Kingdom. Some bogs have preserved bog-wood such as ancient oak logs useful in dendrochronology, and they have yielded extremely well preserved bog bodies, with hair, organs, and skin intact, buried there thousands of years ago after apparent Germanic and Celtic human sacrifice. Excellent examples of such human specimens are Haraldskær Woman and Tollund Man in Denmark, and Lindow man found at Lindow Common in England. At Céide Fields in County Mayo in Ireland, a 5,000-year-old neolithic farming landscape has been found preserved under a blanket bog, complete with field walls and hut sites. One ancient artifact found in bogs in many places is bog butter, large masses of fat, usually in wooden containers. These are thought to have been food stores, of both butter and tallow.

Lake

A lake is an area of variable size filled with water, localized in a basin, that is surrounded by land, apart from any river or other outlet that serves to feed or drain the lake. Lakes lie on land and are not part of the ocean, and therefore are distinct from lagoons, and are also larger and deeper than ponds, though there are no official or scientific definitions. Lakes can be contrasted with rivers or streams, which are usually flowing. Most lakes are fed and drained by rivers and streams.

An area of lakes in Germany at Mecklenburg Lakeland

Natural lakes are generally found in mountainous areas, rift zones, and areas with ongoing glaciation. Other lakes are found in endorheic basins or along the courses of mature rivers. In some parts of the world there are many lakes because of chaotic drainage patterns left over from the last Ice Age. All lakes are temporary over geologic time scales, as they will slowly fill in with sediments or spill out of the basin containing them.

Many lakes are artificial and are constructed for industrial or agricultural use, for hydro-electric power generation or domestic water supply, or for aesthetic or recreational purposes or even for other activities.

Etymology, Meaning, and Usage of "Lake"

Oeschinen Lake in the Swiss Alps

The word *lake* comes from Middle English *lake* ("lake, pond, waterway"), from Old English *lacu* ("pond, pool, stream"), from Proto-Germanic *lakō* ("pond, ditch, slow moving stream"), from the Proto-Indo-European root *leǵ-* ("to leak, drain"). Cognates include Dutch *laak* ("lake, pond, ditch"), Middle Low German *lāke* ("water pooled in a riverbed, puddle") as in: de:Moorlake, de:Wolfslake, de:Butterlake, German *Lache* ("pool, puddle"), and Icelandic *lækur* ("slow flowing stream"). Also related are the English words *leak* and *leach*.

Lake Tahoe on the border of California and Nevada

There is considerable uncertainty about defining the difference between lakes and ponds, and no current internationally accepted definition of either term across scientific disciplines or political boundaries exists. For example, limnologists have defined lakes as water bodies which are simply a larger version of a pond, which can have wave action on the shoreline or where wind-induced turbulence plays a major role in mixing the water column. None of these definitions completely excludes ponds and all are difficult to measure. For this reason, simple size-based definitions are increasingly used to separate ponds and lakes. One definition of *lake* is a body of water of 2 hectares (5 acres) or more in area; however, others have defined lakes as waterbodies of 5 hectares (12 acres) and above, or 8 hectares (20 acres) and above. Charles Elton, one of the founders of ecology, regarded lakes as waterbodies of 40 hectares (99 acres) or more. The term *lake* is also used to describe a feature such as Lake Eyre, which is a dry basin most of the time but may become filled under seasonal conditions of heavy rainfall. In common usage, many lakes bear names ending with the word *pond*, and a lesser number of names ending with *lake* are in quasi-technical fact, ponds. One textbook illustrates this point with the following: "In Newfoundland, for example, almost every lake is called a pond, whereas in Wisconsin, almost every pond is called a lake."

The Caspian Sea is either the world's largest lake or a full-fledged sea.

One hydrology book proposes to define the term "lake" as a body of water with the following five characteristics:

- it partially or totally fills one or several basins connected by straits

- has essentially the same water level in all parts (except for relatively short-lived variations caused by wind, varying ice cover, large inflows, etc.)

- it does not have regular intrusion of seawater

- a considerable portion of the sediment suspended in the water is captured by the basins (for this to happen they need to have a sufficiently small inflow-to-volume ratio)

- the area measured at the mean water level exceeds an arbitrarily chosen threshold (for instance, one hectare)

With the exception of the seawater intrusion criterion, the others have been accepted or elaborated upon by other hydrology publications.

Distribution of Lakes

The Seven Rila Lakes are a group of glacial lakes in the Bulgarian Rila mountains.

The majority of lakes on Earth are fresh water, and most lie in the Northern Hemisphere at higher latitudes. Canada, with a deranged drainage system has an estimated 31,752 lakes larger than 3 square kilometres (1.2 sq mi) and an unknown total number of lakes, but is estimated to be at least 2 million. Finland has 187,888 lakes 500 square metres (5,400 sq ft) or larger, of which 56,000 are large (10,000 square metres (110,000 sq ft) or larger).

Most lakes have at least one natural outflow in the form of a river or stream, which maintain a lake's average level by allowing the drainage of excess water. Some lakes do not have a natural outflow and lose water solely by evaporation or underground seepage or both. They are termed endorheic lakes.

Wayanad District Kerala India

Many lakes are artificial and are constructed for hydro-electric power generation, aesthetic purposes, recreational purposes, industrial use, agricultural use or domestic water supply.

Evidence of extraterrestrial lakes exists; "definitive evidence of lakes filled with methane" was announced by NASA as returned by the Cassini Probe observing the moon Titan, which orbits the planet Saturn.

Globally, lakes are greatly outnumbered by ponds: of an estimated 304 million standing water bodies worldwide, 91% are 1 hectare (2.5 acres) or less in area. Small lakes are also much more numerous than large lakes: in terms of area, one-third of the world's standing water is represented by lakes and ponds of 10 hectares (25 acres) or less. However, large lakes account for much of the area of standing water with 122 large lakes of 1,000 square kilometres (390 sq mi, 100,000 ha, 247,000 acres) or more representing about 29% of the total global area of standing inland water.

Origin of Natural Lakes

A portion of the Great Salt Lake in Utah, United States

Since they progressively become filled by sediment, lakes are considered ephemeral over geological time scales, and long-living lakes imply that active processes keep forming the basins in which they form. There are a number of natural processes that can form lakes.

A lake in the Andes Mountains

Tectonic Lakes

The longest-living lakes on Earth are related to tectonic processes which is created due to a tectonic uplift of a mountain range which create depressions that accumulate water and form lakes. The forming process of tectonic lake is: the earth surface, perhaps the high mountain and plateau, or the hills or plain - there appeared rupture on the ground, and depression along rupture direction - water begins to be stored gradually, and form the lake. Examples of tectonic lakes include Lake Balaton (Central Europe), Holxil Lake, Zonag Lake, and Kusai Lake (In the depression belt between Aerdaijin Mountains and Holxil Mountains in Qinghai Province, China), Xijir Ulan Lake, Ulan Ul Lake, and Duogecuoren Lake (Holxil Mountains and Tanggula Mountains, China), Ngoring Lake and Gyaring Lake (In the upper reaches of the Yellow River, China).

Landslide and Ice-dam Lakes

Lakes can also form by means of landslides or by glacial blockages. An example of the latter occurred during the last ice age in the U.S. state of Washington, when a huge lake formed behind a glacial flow; when the ice retreated, the result was an immense flood that created the Dry Falls at Sun Lakes, Washington.

Salt Lakes

Salt lakes (also called saline lakes) can form where there is no natural outlet or where the water evaporates rapidly and the drainage surface of the water table has a higher-than-normal salt content. Examples of salt lakes include Great Salt Lake, the Aral Sea, and the Dead Sea.

Oxbow Lakes

Small, crescent-shaped lakes called oxbow lakes can form in river valleys as a result of meandering. The slow-moving river forms a sinuous shape as the outer side of bends are eroded away more rapidly than the inner side. Eventually a horseshoe bend is formed and the river cuts through the narrow neck. This new passage then forms the main passage for the river and the ends of the bend become silted up, thus forming a bow-shaped lake.

Crater Lakes

Crater lakes are formed in volcanic craters and calderas which fill up with precipitation more rapidly than they empty via evaporation. Sometimes the latter are called caldera lakes, although often no distinction is made. An example is Crater Lake in Oregon, in the caldera of Mount Mazama. The caldera was created in a massive volcanic eruption that led to the subsidence of Mount Mazama around 4860 BC.

Glacial Lakes

The advance and retreat of glaciers can scrape depressions in the surface where water accumulates; such lakes are common in Scandinavia, Patagonia, Siberia and Canada. The most notable examples are probably the Great Lakes of North America. As a particular case, gloe lakes are basins that have emerged from the sea as a consequence of post-glacial rebound, and are now filled with freshwater.

Other Lakes

Some lakes, such as Lake Jackson in Florida, USA, come into existence as a result of sinkhole activity.

Lake Vostok is a subglacial lake in Antarctica, possibly the largest in the world. The pressure from the ice atop it and its internal chemical composition mean that, if the lake were drilled into, a fissure could result that would spray somewhat like a geyser.

Lake- Yaounde

Most lakes are geologically young and shrinking since the natural results of erosion will tend to wear away the sides and fill the basin. Exceptions are those such as Lake Baikal and Lake Tanganyika that lie along continental rift zones and are created by the crust's subsidence as two plates are pulled apart. These lakes are the oldest and deepest in the world. Lake Baikal, which is 25–30 million years old, is deepening at a faster rate than it is being filled by erosion and may be destined over millions of years to become attached to the global ocean. The Red Sea, for example, is thought to have originated as a rift-valley lake.

Types of Lakes

One of the many artificial lakes in Arizona at sunset.

Lake Parramatta, an artificial lake in Sydney, Australia.

A naturalized former gravel pit lake in northern Croatia.

The crater lake of Volcán Irazú, Costa Rica.

These kettle lakes in Alaska were formed by a retreating glacier.

Ephemeral 'Lake Badwater', a lake only noted after heavy winter and spring rainfall, Badwater Basin, Death Valley National Park.

- Tectonic lake: Owing to the internal force action of the earth's crust, there produce the tectonic lake basins, which store water and transform into tectonic lakes.

- Periglacial lake: Part of the lake's margin is formed by an ice sheet, ice cap or glacier, the ice having obstructed the natural drainage of the land.

- Subglacial lake: A lake which is permanently covered by ice. They can occur under glaciers, ice caps or ice sheets. There are many such lakes, but Lake Vostok in Antarctica is by far the largest. They are kept liquid because the overlying ice acts as a thermal insulator retaining energy introduced to its underside by friction, by water percolating through crevasses, by the pressure from the mass of the ice sheet above or by geothermal heating below.

- Glacial lake: a lake with origins in a melted glacier, such as a kettle lake.

- Artificial lake: A lake created by flooding land behind a dam, called an impoundment or reservoir, by deliberate human excavation, or by the flooding of an excavation incident to a mineral-extraction operation such as an open pit mine or quarry. Some of the world's largest lakes are reservoirs like Hirakud Dam in India.

Ice Melting on Lake Balaton

- Endorheic lake, terminal or closed: A lake which has no significant outflow, either through rivers or underground diffusion. Any water within an endorheic basin leaves the system only through evaporation or seepage. These lakes, such as Lake Eyre in central Australia, the Aral Sea in central Asia, or the Great Salt Lake in the Western United States, are most common in deserts.

- Meromictic lake: A lake which has layers of water which do not intermix. The deepest layer of water in such a lake does not contain any dissolved oxygen. The layers of sediment at the bottom of a meromictic lake remain relatively undisturbed because there are no living aerobic organisms.

- Fjord lake: A lake in a glacially eroded valley that has been eroded below sea level.

- Oxbow lake: A lake which is formed when a wide meander from a stream or a river is cut off to form a lake. They are called "oxbow" lakes due to the distinctive curved shape that results from this process.

- Rift lake or sag pond: A lake which forms as a result of subsidence along a geological fault in the Earth's tectonic plates. Examples include the Rift Valley lakes of eastern Africa and Lake Baikal in Siberia.

- Underground lake: A lake which is formed under the surface of the Earth's crust. Such a lake may be associated with caves, aquifers or springs.

- Crater lake: A lake which forms in a volcanic caldera or crater after the volcano has been inactive for some time. Water in this type of lake may be fresh or highly acidic, and may contain various dissolved minerals. Some also have geothermal activity, especially if the volcano is merely dormant rather than extinct.

- Lava lake: A pool of molten lava contained in a volcanic crater or other depression. Lava lakes that have partly or completely solidified are also referred to as lava lakes.

- Former: A lake which is no longer in existence. Such lakes include prehistoric lakes and lakes which have permanently dried up through evaporation or human intervention. Owens Lake in California, USA, is an example of a former lake. Former lakes are a common feature of the Basin and Range area of south-western North America.

- Ephemeral lake, intermittent lake, or seasonal lake: A seasonal lake that exists as a body of water during part of the year.

- Shrunken: Closely related to *former* lakes, a shrunken lake is one which has drastically decreased in size over geological time. Lake Agassiz, which once covered much of central North America, is a good example of a shrunken lake. Two notable remnants of this lake are Lake Winnipeg and Lake Winnipegosis.

- Eolic lake: A lake which forms in a depression created by the activity of the winds.

- Vlei, in South Africa, shallow lakes which vary considerably with seasons.

- Epishelf lakes, unique lakes which exist on top of a dense saltwater body and are surrounded by ice. These are mostly found in the Antarctica.

Characteristics

Many lakes can have tremendous cultural importance. The West Lake of Hangzhou has inspired romantic poets throughout the ages, and has been an important influence on garden designs in China, Japan and Korea.

Lakes have numerous features in addition to lake type, such as drainage basin (also known as catchment area), inflow and outflow, nutrient content, dissolved oxygen, pollutants, pH, and sedimentation.

Changes in the level of a lake are controlled by the difference between the input and output compared to the total volume of the lake. Significant input sources are precipitation onto the lake, runoff carried by streams and channels from the lake's catchment area, groundwater channels and aquifers, and artificial sources from outside the catchment area. Output sources are evaporation from the lake, surface and groundwater flows, and any extraction of lake water by humans. As climate conditions and human water requirements vary, these will create fluctuations in the lake level.

Lake Päijänne is one of tens of thousands of lakes in Finnish Lakeland.

Lake Mapourika, New Zealand

Lakes can be also categorized on the basis of their richness in nutrients, which typically affect plant growth. Nutrient-poor lakes are said to be *oligotrophic* and are generally clear, having a low concentration of plant life. *Mesotrophic lakes* have good clarity and an average level of nutrients. *Eutrophic* lakes are enriched with nutrients, resulting in good plant growth and possible algal blooms. *Hypertrophic* lakes are bodies of water that have been excessively enriched with nutrients. These lakes typically have poor clarity and are subject to devastating algal blooms. Lakes typically reach this condition due to human activities, such as heavy use of fertilizers in the lake catchment area. Such lakes are of little use to humans and have a poor ecosystem due to decreased dissolved oxygen.

Due to the unusual relationship between water's temperature and its density, lakes form layers called thermoclines, layers of drastically varying temperature relative to depth. Fresh water is most dense at about 4 degrees Celsius (39.2 °F) at sea level. When the temperature of the water at the surface of a lake reaches the same temperature as deeper water, as it does during the cooler months in temperate climates, the water in the lake can mix, bringing oxygen-starved water up from the depths and bringing oxygen down to decomposing sediments. Deep temperate lakes can maintain a reservoir of cold water year-round, which allows some cities to tap that reservoir for deep lake water cooling.

Lake Teletskoye, Siberia

Since the surface water of deep tropical lakes never reaches the temperature of maximum density, there is no process that makes the water mix. The deeper layer becomes oxygen starved and can become saturated with carbon dioxide, or other gases such as sulfur dioxide if there is even a trace of volcanic activity. Exceptional events, such as earthquakes or landslides, can cause mixing which rapidly brings the deep layers up to the surface and release a vast cloud of gas which lay trapped in solution in the colder water at the bottom of the lake. This is called a limnic eruption. An example is the disaster at Lake Nyos in Cameroon. The amount of gas that can be dissolved in water is directly related to pressure. As deep water surfaces, the pressure drops and a vast amount of gas comes out of solution. Under these circumstances carbon dioxide is hazardous because it is heavier than air and displaces it, so it may flow down a river valley to human settlements and cause mass asphyxiation.

The material at the bottom of a lake, or *lake bed*, may be composed of a wide variety of inorganics, such as silt or sand, and organic material, such as decaying plant or animal matter. The composition of the lake bed has a significant impact on the flora and fauna found within the lake's environs by contributing to the amounts and the types of nutrients available.

A paired (black and white) layer of the varved lake sediments correspond to a year. During winter, when organisms die, carbon is deposited down, resulting to a black layer. At the same year, during summer, only few organic materials are deposited, resulting to a white layer at the lake bed. These are commonly used to track past paleontological events.

Natural lakes provide a microcosm of living and nonliving elements that are relatively independent of their surrounding environments. Therefore, lake organisms can often be studied in isolation from the lake's surroundings.

Limnology

Lake Billy Chinook, Deschutes National Forest, Oregon.

Limnology is the study of inland bodies of water and related ecosystems. Limnology divides lakes into three zones: the *littoral zone*, a sloped area close to land; the *photic* or *open-water zone*, where sunlight is abundant; and the deep-water *profundal* or *benthic zone*, where little sunlight can reach. The depth to which light can reach in lakes depends on turbidity, determined by the density and size of suspended particles. A particle is in suspension if its weight is less than the random turbidity forces acting upon it. These particles can be sedimentary or biological in origin and are responsible for the color of the water. Decaying plant matter, for instance, may be responsible for a yellow or brown color, while algae may cause greenish water. In very shallow water bodies, iron oxides make water reddish brown. Biological particles include algae and detritus. Bottom-dwelling detritivorous fish can be responsible for turbid waters, because they stir the mud in search of food. Piscivorous fish contribute to turbidity by eating plant-eating (planktonivorous) fish, thus increasing the amount of algae. The light depth or transparency is measured by using a *Secchi disk*, a 20-cm (8 in) disk with alternating white and black quadrants. The depth at which the disk is no longer visible is the *Secchi depth*, a measure of transparency. The Secchi disk is commonly used to test for eutrophication.

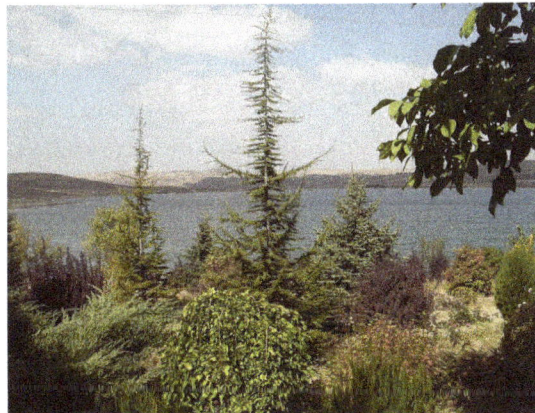

Lake Cugun, Kirsehir, Turkey.

A lake moderates the surrounding region's temperature and climate because water has a very high specific heat capacity (4,186 J·kg⁻¹·K⁻¹). In the daytime a lake can cool the land beside it with local winds, resulting in a sea breeze; in the night it can warm it with a land breeze.

Lake of Flowers (Liqeni i Lulëve), one of the Lurë Mountains glacial lakes, Albania.

How Lakes Disappear

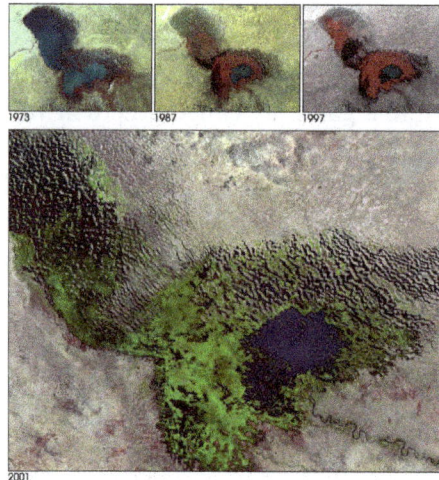

Lake Chad in a 2001 satellite image, with the actual lake in blue, and vegetation on top of the old lake bed in green.

The lake may be infilled with deposited sediment and gradually become a wetland such as a swamp or marsh. Large water plants, typically reeds, accelerate this closing process significantly because they partially decompose to form peat soils that fill the shallows. Conversely, peat soils in a marsh can naturally burn and reverse this process to recreate a shallow lake resulting in a dynamic equilibrium between marsh and lake. This is significant since wildfire has been largely suppressed in the developed world over the past century. This has artificially converted many shallow lakes into emergent marshes. Turbid lakes and lakes with many plant-eating fish tend to disappear more slowly. A "disappearing" lake (barely noticeable on a human timescale) typically has extensive plant mats at the water's edge. These become a new habitat for other plants, like peat moss when conditions are right, and animals, many of which are very rare. Gradually the lake closes and young peat may form, forming a fen. In lowland river valleys where a river can meander, the presence of peat is explained by the infilling of historical ox-bow lakes. In the very last stages of succession, trees can grow in, eventually turning the wetland into a forest.

Lake Badwater, February 9, 2005. Landsat 5 satellite photo.

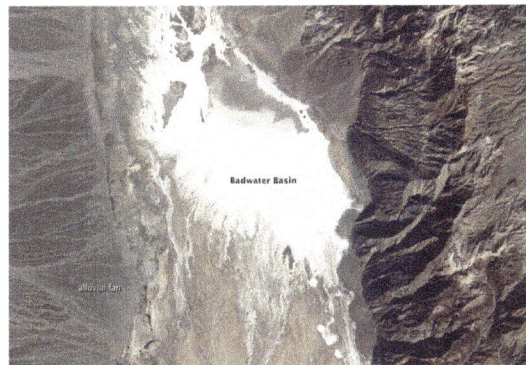

Badwater Basin dry lake, February 15, 2007. Landsat 5 satellite photo.

Some lakes can disappear seasonally. These are called intermittent lakes, ephemeral lakes, or seasonal lakes and can be found in karstic terrain. A prime example of an intermittent lake is Lake Cerknica in Slovenia or Lag Prau Pulte in Graubünden. Other intermittent lakes are only the result of above-average precipitation in a closed, or endorheic basin, usually filling dry lake beds. This can occur in some of the driest places on earth, like Death Valley. This occurred in the spring of 2005, after unusually heavy rains. The lake did not last into the summer, and was quickly evaporated. A more commonly filled lake of this type is Sevier Lake of west-central Utah.

Sometimes a lake will disappear quickly. On 3 June 2005, in Nizhny Novgorod Oblast, Russia, a lake called Lake Beloye vanished in a matter of minutes. News sources reported that government officials theorized that this strange phenomenon may have been caused by a shift in the soil underneath the lake that allowed its water to drain through channels leading to the Oka River.

The presence of ground permafrost is important to the persistence of some lakes. According to research published in the journal *Science* ("Disappearing Arctic Lakes", June 2005), thawing permafrost may explain the shrinking or disappearance of hundreds of large Arctic lakes across western Siberia. The idea here is that rising air and soil temperatures thaw permafrost, allowing the lakes to drain away into the ground.

Some lakes disappear because of human development factors. The shrinking Aral Sea is described as being "murdered" by the diversion for irrigation of the rivers feeding it.

Extraterrestrial Lakes

Only one world other than Earth is known to harbor large lakes, Saturn's largest moon, Titan. Photographs and spectroscopic analysis by the Cassini–Huygens spacecraft show liquid ethane on the surface, which is thought to be mixed with liquid methane. The largest Titanean lake, Kraken Mare at 400,000 km², is three-times the size of any lake on Earth, and even the second, Ligeia Mare, is estimated to be slightly larger than Earth's Lake Michigan–Huron.

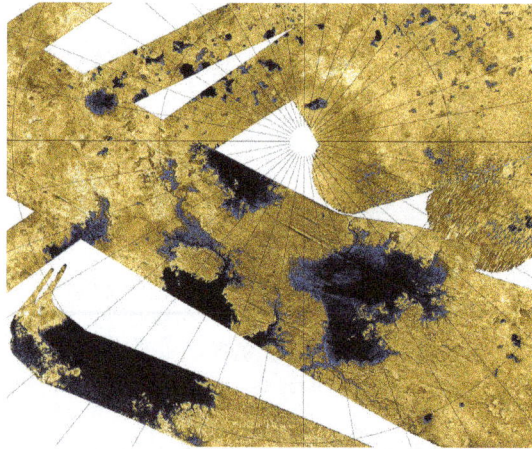

Titan's north polar hydrocarbon seas and lakes as seen in a false-color *Cassini* synthetic aperture radar mosaic.

Jupiter's large moon Io is volcanically active, and as a result sulfur deposits have accumulated on the surface. Some photographs taken during the Galileo mission appear to show lakes of liquid sulfur in volcanic caldera, though these are more analogous to lake of lava than of water on Earth.

The planet Mars is too cold and has too little atmospheric pressure to permit the pooling of liquid water. Geologic evidence appears to confirm, however, that ancient lakes once formed on the surface. It is also possible that volcanic activity on Mars will occasionally melt subsurface ice, creating large temporary lakes. This water would quickly freeze and then sublimate, unless insulated in some manner, such as by a coating of volcanic ash.

There are dark basaltic plains on the Moon, similar to lunar maria but smaller, that are called *lacus* (singular *lacus*, Latin for "lake") because they were thought by early astronomers to be lakes of water.

Notable Lakes on Earth

Round Tangle Lake, one of the Tangle Lakes, 2,864 feet (873 m) above sea level in interior Alaska

- The largest lake by surface area is Lake Michigan-Huron, which is hydrologically a single lake. Its surface area is 45,300 sq. mi./117,400 km². For those who consider Lake Michigan-Huron to be separate lakes, Lake Superior would be the largest at 31,700 sq. mi./82,100 km².

- The deepest lake is Lake Baikal in Siberia, with a bottom at 1,637 metres (5,371 ft). Its mean depth is also the greatest in the world (749 metres (2,457 ft)). It is also the world's largest lake by volume (23,600 cubic kilometres (5,700 cu mi), though smaller than the Caspian Sea at 78,200 cubic kilometres (18,800 cu mi)), and the second longest (about 630 kilometres (390 mi) from tip to tip).

- The longest lake is Lake Tanganyika, with a length of about 660 kilometres (410 mi) (measured along the lake's center line). It is also the second largest by volume and second deepest (1,470 metres (4,820 ft)) in the world, after lake Baikal.

- The world's oldest lake is Lake Baikal, followed by Lake Tanganyika (Tanzania). Lake Maracaibo is considered by some to be the second-oldest lake on Earth, but since it lies at sea level and nowadays is a contiguous body of water with the sea, others consider that it has turned into a small bay.

- The world's highest lake, if size is not a criterion, may be the crater lake of Ojos del Salado, at 6,390 metres (20,965 ft).

- The highest large (greater than 250 square kilometres (97 sq mi)) lake in the world is the 290 square kilometres (110 sq mi) Pumoyong Tso (Pumuoyong

Tso), in the Tibet Autonomous Region of China, at 28-34N 90-24E, 5,018 metres (16,463 ft) above sea level.

- The world's highest commercially navigable lake is Lake Titicaca in Peru and Bolivia at 3,812 m (12,507 ft). It is also the largest lake in South America.

- The world's lowest lake is the Dead Sea, bordering Israel and Jordan at 418 metres (1,371 ft) below sea level. It is also one of the lakes with highest salt concentration.

- Lake Michigan–Huron has the longest lake coastline in the world: about 5,250 kilometres (3,260 mi), excluding the coastline of its many inner islands. Even if it is considered two lakes, Lake Huron alone would still have the longest coastline in the world at 2,980 kilometres (1,850 mi).

- The largest island in a lake is Manitoulin Island in Lake Huron, with a surface area of 2,766 square kilometres (1,068 sq mi). Lake Manitou, on Manitoulin Island, is the largest lake on an island in a lake.

- The largest lake on an island is Nettilling Lake on Baffin Island, with an area of 5,542 square kilometres (2,140 sq mi) and a maximum length of 123 kilometres (76 mi).

- The largest lake in the world that drains naturally in two directions is Wollaston Lake.

- Lake Toba on the island of Sumatra is in what is probably the largest resurgent caldera on Earth.

- The largest lake completely within the boundaries of a single city is Lake Wanapitei in the city of Sudbury, Ontario, Canada. Before the current city boundaries came into effect in 2001, this status was held by Lake Ramsey, also in Sudbury.

- Lake Enriquillo in Dominican Republic is the only saltwater lake in the world inhabited by crocodiles.

- Lake Bernard, Ontario, Canada, claims to be the largest lake in the world with no islands.

- The largest lake in one country is Lake Michigan, in the U.S.A. However, it is sometimes considered part of Lake Michigan-Huron, making the record go to Great Bear Lake, Northwest Territories, in Canada, the largest lake within one jurisdiction.

- The largest lake on an island in a lake on an island is Crater Lake on Vulcano Island in Lake Taal on the island of Luzon, The Philippines.

- The northernmost named lake on Earth is Upper Dumbell Lake in the Qikiqtaaluk Region of Nunavut, Canada at a latitude of 82°28'N. It is 5.2 kilometres (3.2 mi) southwest of Alert, the northernmost settlement in the world. There are also several small lakes north of Upper Dumbell Lake, but they are all unnamed and only appear on very detailed maps.

Largest by Continent

The largest lakes (surface area) by continent are:

- Australia – Lake Eyre (salt lake)

- Africa – Lake Victoria, also the third-largest freshwater lake on Earth. It is one of the Great Lakes of Africa.

- Antarctica – Lake Vostok (subglacial)

- Asia – Lake Baikal (if the Caspian Sea is considered a lake, it is the largest in Eurasia, but is divided between the two geographic continents)

- Oceania – Lake Eyre when filled; the largest permanent (and freshwater) lake in Oceania is Lake Taupo.

- Europe – Lake Ladoga, followed by Lake Onega, both in northwestern Russia.

- North America – Lake Michigan-Huron, which is hydrologically a single lake. However, lakes Huron and Michigan are usually considered separate lakes, in which case Lake Superior would be the largest.

- South America – Lake Titicaca, which is also the highest navigable body of water on Earth at 3,812 metres (12,507 ft) above sea level. The much larger Lake Maracaibo is much older, but perceived by some to no longer be genuinely a lake for multiple reasons.

River

A river is a natural flowing watercourse, usually freshwater, flowing towards an ocean, sea, lake or another river. In some cases a river flows into the ground and becomes dry at the end of its course without reaching another body of water. Small rivers can be referred to using names such as stream, creek, brook, rivulet, and rill. There are no official definitions for the generic term river as applied to geographic features, although in some countries or communities a stream is defined by its size. Many names for small rivers are specific to geographic location; examples are "run" in some parts of the United States, "burn" in Scotland and northeast England, and "beck" in northern England. Sometimes a river is defined as being larger than a creek, but not always: the language is vague.

Rivers are part of the hydrological cycle. Water generally collects in a river from precipitation through a drainage basin from surface runoff and other sources such as groundwater recharge, springs, and the release of stored water in natural ice and snowpacks (e.g. from glaciers). Potamology is the scientific study of rivers while limnology is the study of inland waters in general.

Extraterrestrial rivers of liquid hydrocarbons have recently been found on Titan. Channels may indicate past rivers on other planets, specifically outflow channels on Mars and rivers are theorised to exist on planets and moons in habitable zones of stars.

Topography

Melting toe of Athabasca Glacier, Jasper National Park, Alberta, Canada

A river begins at a source (or more often several sources), follows a path called a course, and ends at a mouth or mouths. The water in a river is usually confined to a channel, made up of a stream bed between banks. In larger rivers there is often also a wider floodplain shaped by flood-waters over-topping the channel. Floodplains may be very wide in relation to the size of the river channel. This distinction between river channel and floodplain can be blurred, especially in urban areas where the floodplain of a river channel can become greatly developed by housing and industry.

The Loboc River in Bohol, Philippines

Rivers can flow down mountains, through valleys (depressions) or along plains, and can create canyons or gorges.

The term upriver (or upstream) refers to the direction towards the source of the river, i.e. against the direction of flow. Likewise, the term downriver (or downstream) describes the direction towards the mouth of the river, in which the current flows.

The Colorado River at Horseshoe Bend, Arizona

The term left bank refers to the left bank in the direction of flow, right bank to the right.

The river channel typically contains a single stream of water, but some rivers flow as several interconnecting streams of water, producing a braided river. Extensive braided rivers are now found in only a few regions worldwide, such as the South Island of New Zealand. They also occur on peneplains and some of the larger river deltas. Anastamosing rivers are similar to braided rivers and are also quite rare. They have multiple sinuous channels carrying large volumes of sediment. There are rare cases of river bifurcation in which a river divides and the resultant flows ending in different seas. An example is the bifurcation of Nerodime River in Kosovo.

The River Cam from the Green Dragon Bridge, Cambridge (Britain)

A river flowing in its channel is a source of energy which acts on the river channel to change its shape and form. In 1757, the German hydrologist Albert Brahms empirically observed that the submerged weight of objects that may be carried away by a river is proportional to the sixth power of the river flow speed. This formulation is also sometimes called Airy's law. Thus, if the speed of flow is doubled, the flow would dislodge objects with 64 times as much submerged weight. In mountainous torrential zones this can be seen as erosion channels through hard rocks and the creation of sands and gravels from the destruction of larger rocks. In U-shaped glaciated valleys, the subsequent river valley can often easily be identified by the V-shaped channel that it has carved. In the middle reaches where a river flows over flatter land, meanders may form through erosion of the river banks and deposition on the inside of bends. Sometimes the river will cut off a loop, shortening the channel and forming an oxbow lake or billabong. Rivers that carry large amounts of sediment may develop conspicuous deltas at their mouths. Rivers whose mouths are in saline tidal waters may form estuaries.

Throughout the course of the river, the total volume of water transported downstream will often be a combination of the free water flow together with a substantial volume flowing through sub-surface rocks and gravels that underlie the river and its floodplain (called the hyporheic zone). For many rivers in large valleys, this unseen component of flow may greatly exceed the visible flow.

Subsurface Streams

Most but not all rivers flow on the surface. Subterranean rivers flow underground in caves or caverns. Such rivers are frequently found in regions with limestone geologic formations. Subglacial streams are the braided rivers that flow at the beds of glaciers and ice sheets, permitting meltwater to be discharged at the front of the glacier. Because of the gradient in pressure due to the overlying weight of the glacier, such streams can even flow uphill.

Permanence of Flow

An *intermittent river* (or *ephemeral river*) only flows occasionally and can be dry for several years at a time. These rivers are found in regions with limited or highly variable rainfall, or can occur because of geologic conditions such as a highly permeable river bed. Some ephemeral rivers flow during the summer months but not in the winter. Such rivers are typically fed from chalk aquifers which recharge from winter rainfall. In England these rivers are called *bournes* and give their name to places such as Bournemouth and Eastbourne. Even in humid regions, the location where flow begins in the smallest tributary streams generally moves upstream in response to precipitation and downstream in its absence or when active summer vegetation diverts water for evapotranspiration. Normally-dry rivers in arid zones are often identified as arroyos or other regional names.

The meltwater from large hailstorms can create a slurry of water, hail and sand or soil, forming temporary rivers.

Classification

Nile River delta, as seen from Earth orbit. The Nile is an example of a wave-dominated delta that has the classic Greek letter delta (Δ) shape after which river deltas were named.

A radar image of a 400-kilometre (250 mi) river of methane and ethane near the north pole of Saturn's moon Titan

Rivers have been classified by many criteria including their topography, their biotic status, and their relevance to white water rafting or canoeing activities.

Topographical Classification

Rivers can generally be classified as either alluvial, bedrock, or some mix of the two. Alluvial rivers have channels and floodplains that are self-formed in unconsolidated or weakly consolidated sediments. They erode their banks and deposit material on bars and their floodplains. Bedrock rivers form when the river downcuts through the modern sediments and into the underlying bedrock. This occurs in regions that have experienced some kind of uplift (thereby steepening river gradients) or in which a particular hard lithology causes a river to have a steepened reach that has not been covered in modern alluvium. Bedrock rivers very often contain alluvium on their beds; this material is important in eroding and sculpting the channel. Rivers that go through patches of bedrock and patches of deep alluvial cover are classified as mixed bedrock-alluvial.

Alluvial rivers can be further classified by their channel pattern as meandering, braided, wandering, anastomose, or straight. The morphology of an alluvial river reach is controlled by a combination of sediment supply, substrate composition, discharge, vegetation, and bed aggradation.

At the start of the 20th century William Morris Davis devised the "cycle of erosion" method of classifying rivers based on their "age". Although Davis's system is still found in many books today, after the 1950s and 1960s it became increasingly criticized and rejected by geomorphologists. His scheme did not produce testable hypotheses and was therefore deemed non-scientific. Examples of Davis's river "ages" include:

- Youthful river: A river with a steep gradient that has very few tributaries and flows quickly. Its channels erode deeper rather than wider. Examples are the Brazos, Trinity and Ebro rivers.

- Mature river: A river with a gradient that is less steep than those of youthful rivers and flows more slowly. A mature river is fed by many tributaries and has more discharge than a youthful river. Its channels erode wider rather than deeper. Examples are the Mississippi, Saint Lawrence, Danube, Ohio, Thames and Paraná rivers.

- Old river: A river with a low gradient and low erosive energy. Old rivers are characterized by flood plains. Examples are the Yellow, lower Ganges, Tigris, Euphrates, Indus and lower Nile rivers.

- Rejuvenated river: A river with a gradient that is raised by tectonic uplift. Examples are the Rio Grande and Colorado River.

The ways in which a river's characteristics vary between its upper and lower course are summarized by the Bradshaw model. Power-law relationships between channel slope, depth, and width are given as a function of discharge by "river regime".

Biotic Classification

There are several systems of classification based on biotic conditions typically assigning classes from the most oligotrophic or unpolluted through to the most eutrophic or polluted. Other systems are based on a whole eco-system approach such as developed by the New Zealand Ministry for the Environment. In Europe, the requirements of the Water Framework Directive has led to the development of a wide range of classification methods including classifications based on fishery status A system of river zonation used in francophone communities divides rivers into three primary zones:

- The *crenon* is the uppermost zone at the source of the river. It is further divided into the eucrenon (spring or boil zone) and the hypocrenon (brook or headstream zone). These areas have low temperatures, reduced oxygen content and slow moving water.

- The *rhithron* is the upstream portion of the river that follows the crenon. It has relatively cool temperatures, high oxygen levels, and fast, turbulent, swift flow.

- The *potamon* is the remaining downstream stretch of river. It has warmer temperatures, lower oxygen levels, slow flow and sandier bottoms.

Whitewater Classification

The International Scale of River Difficulty is used to rate the challenges of navigation—particularly those with rapids. Class I is the easiest and Class VI is the hardest.

Stream Order Classification

The Strahler Stream Order ranks rivers based on the connectivity and hierarchy of contributing tributaries. Headwaters are first order while the Amazon River is twelfth order. Approximately 80% of the rivers and streams in the world are of the first and second order.

In certain languages, distinctions are made among rivers based on their stream order. In French, for example, rivers that run to the sea are called *fleuve*, while other rivers are called *rivière*. For example, in Canada, the Churchill River in Manitoba is called *la rivière Churchill* as it runs to Hudson Bay, but the Churchill River in Labrador is called *le fleuve Churchill* as it runs to the Atlantic Ocean. As most rivers in France are known by their names only without the word *rivière* or *fleuve* (e.g. *la Seine*, not *le fleuve Seine*, even though the Seine is classed as a *fleuve*), one of the most prominent rivers in the Francophonie commonly known as *fleuve* is *le fleuve Saint-Laurent* (the Saint Lawrence River).

Since many *fleuves* are large and prominent, receiving many tributaries, the word is sometimes used to refer to certain large rivers that flow into other *fleuves*; however, even small streams that run to the sea are called *fleuve* (e.g. *fleuve côtier*, "coastal *fleuve*").

Uses

Leisure activities on the River Avon at Avon Valley Country Park, Keynsham, United Kingdom. A boat giving trips to the public passes a moored private boat.

Rivers have been used as a source of water, for obtaining food, for transport, as a defensive measure, as a source of hydropower to drive machinery, for bathing, and as a means of disposing of waste.

Rivers have been used for navigation for thousands of years. The earliest evidence of navigation is found in the Indus Valley Civilization, which existed in northwestern India around 3300 BC. Riverine navigation provides a cheap means of transport, and is still used extensively on most major rivers of the world like the Amazon, the Ganges, the Nile, the Mississippi, and the Indus. Since river boats are often not regulated, they contribute a large amount to global greenhouse gas emissions, and to local cancer due to inhaling of particulates emitted by the transports.

In some heavily forested regions such as Scandinavia and Canada, lumberjacks use the river to float felled trees downstream to lumber camps for further processing, saving much effort and cost by transporting the huge heavy logs by natural means.

Rivers have been a source of food since pre-history. They are often a rich source of fish and other edible aquatic life, and are a major source of fresh water, which can be used for drinking and irrigation. Most of the major cities of the world are situated on the banks of rivers. Rivers help to determine the urban form of cities and neighbourhoods and their corridors often present opportunities for urban renewal through the development of foreshoreways such as river walks. Rivers also provide an easy means of disposing of waste water and, in much of the less developed world, other wastes.

Watermill in Belgium

Fast flowing rivers and waterfalls are widely used as sources of energy, via watermills and hydroelectric plants. Evidence of watermills shows them in use for many hundreds of years, for instance in Orkney at Dounby Click Mill. Prior to the invention of steam power, watermills for grinding cereals and for processing wool and other textiles were common across Europe. In the 1890s the first machines to generate power from river water were established at places such as Cragside in Northumberland and in recent decades there has been a significant increase in the development of large scale power generation from water, especially in wet mountainous regions such as Norway.

The coarse sediments, gravel, and sand, generated and moved by rivers are extensively used in construction. In parts of the world this can generate extensive new lake habitats as gravel pits re-fill with water. In other circumstances it can destabilise the river bed and the course of the river and cause severe damage to spawning fish populations which rely on stable gravel formations for egg laying.

In upland rivers, rapids with whitewater or even waterfalls occur. Rapids are often used for recreation, such as whitewater kayaking.

Rivers have been important in determining political boundaries and defending countries. For example, the Danube was a long-standing border of the Roman Empire, and today it forms most of the border between Bulgaria and Romania. The Mississippi in North America and the Rhine in Europe are major east-west boundaries in those continents. The Orange and Limpopo Rivers in southern Africa form the boundaries between provinces and countries along their routes.

Ecosystem

The organisms in the riparian zone respond to changes in river channel location and patterns of flow. The ecosystem of rivers is generally described by the river continuum concept, which has some additions and refinements to allow for dams and waterfalls and temporary extensive flooding. The concept describes the river as a system in which the physical parameters, the availability of food particles and the composition of the ecosystem are continuously changing along its length. The food (energy) that remains from the upstream part is used downstream.

The general pattern is that the first order streams contain particulate matter (decaying leaves from the surrounding forests) which is processed there by shredders like Plecoptera larvae. The products of these shredders are used by collectors, such as Hydropsychidae, and further downstream algae that create the primary production become the main food source of the organisms. All changes are gradual and the distribution of each species can be described as a normal curve, with the highest density where the conditions are optimal. In rivers succession is virtually absent and the composition of the ecosystem stays fixed in time.

Chemistry

The chemistry of rivers is complex and depends on inputs from the atmosphere, the geology through which it travels and the inputs from man's activities. The chemical composition of the water has a large impact on the ecology of that water for both plants and animals and it also affects the uses that may be made of the river water. Understanding and characterising river water chemistry requires a well designed and managed sampling and analysis.

Brackish Water

Some rivers generate brackish water by having their river mouth in the ocean. This, in effect creates a unique environment in which certain species are found.

Flooding

Flash flooding caused by a large amount of rain falling in a short amount of time

Flooding is a natural part of a river's cycle. The majority of the erosion of river channels and the erosion and deposition on the associated floodplains occur during the flood stage. In many developed areas, human activity has changed the form of river channels, altering magnitudes and frequencies of flooding. Some examples of this are the building of levees, the straightening of channels, and the draining of natural wetlands. In many cases human activities in rivers and floodplains have dramatically increased the risk of flooding. Straightening rivers allows water to flow more rapidly downstream, increasing the risk of flooding places further downstream. Building on flood plains removes flood storage, which again exacerbates downstream flooding. The building of levees only protects the area behind the levees and not those further downstream. Levees and flood-banks can also increase flooding upstream because of the back-water pressure as the river flow is impeded by the narrow channel banks.

Flow

Studying the flows of rivers is one aspect of hydrology.

Direction

Rivers flow downhill with their power derived from gravity. The direction can involve all directions of the compass and can be a complex meandering path.

Rivers flowing downhill, from river source to river mouth, do not necessarily take the shortest path. For alluvial streams, straight and braided rivers have very low sinuosity and flow directly down hill, while meandering rivers flow from side to side across a valley. Bedrock rivers typically flow in either a fractal pattern, or a pattern that is determined by weaknesses in the bedrock, such as faults, fractures, or more erodible layers.

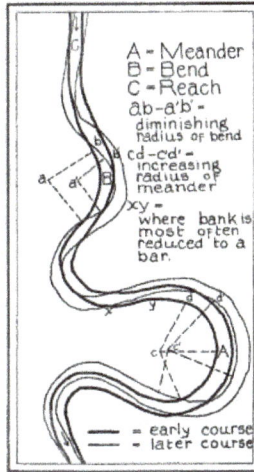

River meandering course

Rate

Volumetric flow rate, also known as discharge, volume flow rate, and rate of water flow, is the volume of water which passes through a given cross-section of the river channel per unit time. It is typically measured in cubic metres per second (cumec) or cubic feet per second (cfs), where $1 \text{ m}^3/\text{s} = 35.51 \text{ ft}^3/\text{s}$; it is sometimes also measured in litres or gallons per second.

Volumetric flow rate can be thought of as the mean velocity of the flow through a given cross-section, times that cross-sectional area. Mean velocity can be approximated through the use of the Law of the Wall. In general, velocity increases with the depth (or hydraulic radius) and slope of the river channel, while the cross-sectional area scales with the depth and the width: the double-counting of depth shows the importance of this variable in determining the discharge through the channel.

Fluvial Erosion

In the youthful stage;

V-shaped valleys: example. River Liffey, Dublin, Ireland.

When the river is subject to vertical erosion, deepening the valley. Hydraulic action loosens and dislodges the rock. The rivers load further erodes its banks and the river bed. Over time, this will deepen the river bed and create steeper sides which are then weathered.

The steepened nature of the banks causes the sides of the valley to move downslope causing the valley to become V-Shaped.

Waterfalls also form in the youthful river valley. example. Powerscourt Waterfall, County Wicklow, Ireland.

Waterfalls usually form where a band of hard rock lies next to a layer of soft rock (easier to erode). Differential erosion occurs as the river can erode the soft rock easier than the hard rock, this leaves the hard rock more elevated and stands out from the river below. Hydraulic action and abrasion are what erodes the soft rock and the water to fall down to the river bed. A plunge pool forms at the bottom and deepens as a result of hydraulic action and abrasion.

Sediment Yield

Frozen river in Alaska

Sediment yield is the total quantity of particulate matter (suspended or bedload) reaching the outlet of a drainage basin over a fixed time frame. Yield is usually expressed as kilograms per square kilometre per year. Sediment delivery processes are affected by a myriad of factors such as drainage area size, basin slope, climate, sediment type (lithology), vegetation cover, and human land use / management practices. The theoretical concept of the 'sediment delivery ratio' (ratio between yield and total amount of sediment eroded) captures the fact that not all of the sediment is eroded within a certain catchment that reaches out to the outlet (due to, for example, deposition on floodplains). Such storage opportunities are typically increased in catchments of larger size, thus leading to a lower yield and sediment delivery ratio.

Management

River bank repair

Rivers are often managed or controlled to make them more useful, or less disruptive, to human activity.

- Dams or weirs may be built to control the flow, store water, or extract energy.

- Levees, known as dikes in Europe, may be built to prevent river water from flowing on floodplains or floodways.

- Canals connect rivers to one another for water transfer or navigation.

- River courses may be modified to improve navigation, or straightened to increase the flow rate.

River management is a continuous activity as rivers tend to 'undo' the modifications made by people. Dredged channels silt up, sluice mechanisms deteriorate with age, levees and dams may suffer seepage or catastrophic failure. The benefits sought through managing rivers may often be offset by the social and economic costs of mitigating the bad effects of such management. As an example, in parts of the developed world, rivers have been confined within channels to free up flat flood-plain land for development. Floods can inundate such development at high financial cost and often with loss of life.

Rivers are increasingly managed for habitat conservation, as they are critical for many aquatic and riparian plants, resident and migratory fishes, waterfowl, birds of prey, migrating birds, and many mammals.

Groundwater

The entire surface water flow of the Alapaha River near Jennings, Florida
going into a sinkhole leading to the Floridan Aquifer groundwater

Groundwater (or ground water) is the water present beneath Earth's surface in soil pore spaces and in the fractures of rock formations. A unit of rock or an unconsolidated deposit is called an aquifer when it can yield a usable quantity of water. The depth at which soil pore spaces or fractures and voids in rock become completely saturated with water is called the water table. Groundwater is recharged from, and eventually flows to, the surface naturally; natural discharge often occurs at springs and seeps, and can form oases or wetlands. Groundwater is also often withdrawn for agricultural, municipal, and industrial use by constructing and operating extraction wells. The study of the distribution and movement of groundwater is hydrogeology, also called groundwater hydrology.

Typically, groundwater is thought of as water flowing through shallow aquifers, but, in the technical sense, it can also contain soil moisture, permafrost (frozen soil), immobile water in very low permeability bedrock, and deep geothermal or oil formation water. Groundwater is hypothesized to provide lubrication that can possibly influence the movement of faults. It is likely that much of Earth's subsurface contains some water, which may be mixed with other fluids in some instances. Groundwater may not be confined only to Earth. The formation of some of the landforms observed on Mars may have been influenced by groundwater. There is also evidence that liquid water may also exist in the subsurface of Jupiter's moon Europa.

Groundwater is often cheaper, more convenient and less vulnerable to pollution than surface water. Therefore, it is commonly used for public water supplies. For example, groundwater provides the largest source of usable water storage in the United States, and California annually withdraws the largest amount of groundwater of all the states. Underground reservoirs contain far more water than the capacity of all surface reservoirs and lakes in the US, including the Great Lakes. Many municipal water supplies are derived solely from groundwater.

Polluted groundwater is less visible, but more difficult to clean up, than pollution in rivers and lakes. Groundwater pollution most often results from improper disposal of wastes on land. Major sources include industrial and household chemicals and garbage landfills, excessive fertilizers and pesticides used in agriculture, industrial waste lagoons, tailings and process wastewater from mines, industrial fracking, oil field brine pits, leaking underground oil storage tanks and pipelines, sewage sludge and septic systems.

Aquifers

An *aquifer* is a layer of porous substrate that contains and transmits groundwater. When water can flow directly between the surface and the saturated zone of an aquifer, the aquifer is unconfined. The deeper parts of unconfined aquifers are usually more saturated since gravity causes water to flow downward.

The upper level of this saturated layer of an unconfined aquifer is called the *water table* or *phreatic surface*. Below the water table, where in general all pore spaces are saturated with water, is the phreatic zone.

Substrate with low porosity that permits limited transmission of groundwater is known as an *aquitard*. An *aquiclude* is a substrate with porosity that is so low it is virtually impermeable to groundwater.

A *confined aquifer* is an aquifer that is overlain by a relatively impermeable layer of rock or substrate such as an aquiclude or aquitard. If a confined aquifer follows a downward grade from its *recharge zone*, groundwater can become pressurized as it flows. This can create artesian wells that flow freely without the need of a pump and rise to a higher elevation than the static water table at the above, unconfined, aquifer.

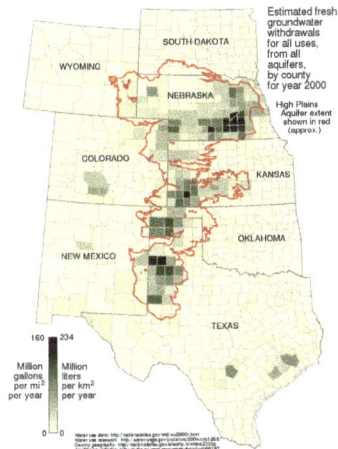

Groundwater withdrawal rates from the Ogallala Aquifer in the Central United States

The characteristics of aquifers vary with the geology and structure of the substrate and topography in which they occur. In general, the more productive aquifers occur in sedimentary geologic formations. By comparison, weathered and fractured crystalline rocks yield smaller quantities of groundwater in many environments. Unconsolidated to poorly cemented alluvial materials that have accumulated as valley-filling sediments in major river valleys and geologically subsiding structural basins are included among the most productive sources of groundwater.

The high specific heat capacity of water and the insulating effect of soil and rock can mitigate the effects of climate and maintain groundwater at a relatively steady temperature. In some places where groundwater temperatures are maintained by this effect at about 10 °C (50 °F), groundwater can be used for controlling the temperature inside structures at the surface. For example, during hot weather relatively cool groundwater can be pumped through radiators in a home and then returned to the ground in another well. During cold seasons, because it is relatively warm, the water can be used in the same way as a source of heat for heat pumps that is much more efficient than using air.

The volume of groundwater in an aquifer can be estimated by measuring water levels in local wells and by examining geologic records from well-drilling to determine the extent, depth and thickness of water-bearing sediments and rocks. Before an investment is made in production wells, test wells may be drilled to measure the depths at which water is encountered and collect samples of soils, rock and water for laboratory analyses. Pumping tests can be performed in test wells to determine flow characteristics of the aquifer.

Water Cycle

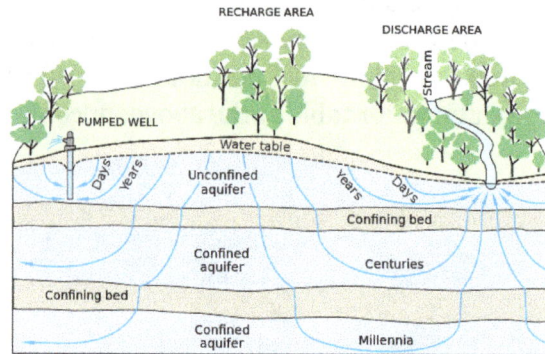

Relative groundwater travel times

Groundwater makes up about twenty percent of the world's fresh water supply, which is about 0.61% of the entire world's water, including oceans and permanent ice. Global groundwater storage is roughly equal to the total amount of freshwater stored in the snow and ice pack, including the north and south poles. This makes it an important resource that can act as a natural storage that can buffer against shortages of surface water, as in during times of drought.

Dzherelo, a common source of drinking water in a Ukrainian village

Groundwater is naturally replenished by surface water from precipitation, streams, and rivers when this recharge reaches the water table.

Groundwater can be a long-term 'reservoir' of the natural water cycle (with residence times from days to millennia), as opposed to short-term water reservoirs like the atmosphere and fresh surface water (which have residence times from minutes to years). The figure shows how deep groundwater (which is quite distant from the surface recharge) can take a very long time to complete its natural cycle.

The Great Artesian Basin in central and eastern Australia is one of the largest confined aquifer systems in the world, extending for almost 2 million km². By analysing the trace elements in water sourced from deep underground, hydrogeologists have been able to determine that water extracted from these aquifers can be more than 1 million years old.

By comparing the age of groundwater obtained from different parts of the Great Artesian Basin, hydrogeologists have found it increases in age across the basin. Where water recharges the aquifers along the Eastern Divide, ages are young. As groundwater flows westward across the continent, it increases in age, with the oldest groundwater occurring in the western parts. This means that in order to have travelled almost 1000 km from the source of recharge in 1 million years, the groundwater flowing through the Great Artesian Basin travels at an average rate of about 1 metre per year.

Reflective carpet trapping soil water vapor

Recent research has demonstrated that evaporation of groundwater can play a significant role in the local water cycle, especially in arid regions. Scientists in Saudi Arabia have proposed plans to recapture and recycle this evaporative moisture for crop irrigation. In the opposite photo, a 50-centimeter-square reflective carpet, made of small adjacent plastic cones, was placed in a plant-free dry desert area for five months, without rain or irrigation. It managed to capture and condense enough ground vapor to bring to life naturally buried seeds underneath it, with a green area of about 10% of the carpet area. It is expected that, if seeds were put down before placing this carpet, a much wider area would become green.

Issues

Overview

Certain problems have beset the use of groundwater around the world. Just as river waters have been over-used and polluted in many parts of the world, so too have aquifers. The big difference is that aquifers are out of sight. The other major problem is that water management agencies, when calculating the "sustainable yield" of aquifer and river water, have often counted the same water twice, once in the aquifer, and once in its connected river. This problem, although understood for centuries, has persisted, partly through inertia within government agencies. In Australia, for example, prior to the statutory reforms initiated by the Council of Australian Governments water reform framework in the 1990s, many Australian states managed groundwater and surface water through separate government agencies, an approach beset by rivalry and poor communication.

In general, the time lags inherent in the dynamic response of groundwater to development have been ignored by water management agencies, decades after scientific understanding of the issue was consolidated. In brief, the effects of groundwater overdraft (although undeniably real) may take decades or centuries to manifest themselves. In a classic study in 1982, Bredehoeft and colleagues modeled a situation where groundwater extraction in an intermontane basin withdrew the entire annual recharge, leaving 'nothing' for the natural groundwater-dependent vegetation community. Even when the borefield was situated close to the vegetation, 30% of the original vegetation demand could still be met by the lag inherent in the system after 100 years. By year 500, this had reduced to 0%, signalling complete death of the groundwater-dependent vegetation. The science has been available to make these calculations for decades; however, in general water management agencies have ignored effects that will appear outside the rough timeframe of political elections (3 to 5 years). Marios Sophocleous argued strongly that management agencies must define and use appropriate timeframes in groundwater planning. This will mean calculating groundwater withdrawal permits based on predicted effects decades, sometimes centuries in the future.

As water moves through the landscape, it collects soluble salts, mainly sodium chloride. Where such water enters the atmosphere through evapotranspiration, these salts are left behind. In irrigation districts, poor drainage of soils and surface aquifers can result in water tables' coming to the surface in low-lying areas. Major land degradation problems of soil salinity and waterlogging result, combined with increasing levels of salt in surface waters. As a consequence, major damage has occurred to local economies and environments.

Four important effects are worthy of brief mention. First, flood mitigation schemes, intended to protect infrastructure built on floodplains, have had the unintended consequence of reducing aquifer recharge associated with natural flooding. Second, prolonged depletion of groundwater in extensive aquifers can result in land subsidence, with associated infrastructure damage – as well as, third, saline intrusion. Fourth, draining acid sulphate soils, often found in low-lying coastal plains, can result in acidification and pollution of formerly freshwater and estuarine streams.

Another cause for concern is that groundwater drawdown from over-allocated aquifers has the potential to cause severe damage to both terrestrial and aquatic ecosystems – in some cases very conspicuously but in others quite imperceptibly because of the extended period over which the damage occurs.

Overdraft

Wetlands contrast the arid landscape around Middle Spring, Fish Springs National Wildlife Refuge, Utah

Groundwater is a highly useful and often abundant resource. However, over-use, or overdraft, can cause major problems to human users and to the environment. The most evident problem (as far as human groundwater use is concerned) is a lowering of the water table beyond the reach of existing wells. As a consequence, wells must be drilled deeper to reach the groundwater; in some places (e.g., California, Texas, and India) the water table has dropped hundreds of feet because of extensive well pumping. In the Punjab region of India, for example, groundwater levels have dropped 10 meters since 1979, and the rate of depletion is accelerating. A lowered water table may, in turn, cause other problems such as groundwater-related subsidence and saltwater intrusion.

Groundwater is also ecologically important. The importance of groundwater to ecosystems is often overlooked, even by freshwater biologists and ecologists. Groundwaters sustain rivers, wetlands, and lakes, as well as subterranean ecosystems within karst or alluvial aquifers.

Not all ecosystems need groundwater, of course. Some terrestrial ecosystems – for example, those of the open deserts and similar arid environments – exist on irregular rainfall and the moisture it delivers to the soil, supplemented by moisture in the air. While there are other terrestrial ecosystems in more hospitable environments where groundwater plays no central role, groundwater is in fact fundamental to many of the world's major ecosystems. Water flows between groundwaters and surface waters. Most rivers, lakes, and wetlands are fed by, and (at other places or times) feed groundwater, to varying degrees. Groundwater feeds soil moisture through percolation, and many terrestrial vegetation communities depend directly on either groundwater or the percolated soil moisture above the aquifer for at least part of each year. Hyporheic zones (the mixing zone of streamwater and groundwater) and riparian zones are examples of ecotones largely or totally dependent on groundwater.

Subsidence

Subsidence occurs when too much water is pumped out from underground, deflating the space below the above-surface, and thus causing the ground to collapse. The result can look like craters on plots of land. This occurs because, in its natural equilibrium state, the hydraulic pressure of groundwater in the pore spaces of the aquifer and the aquitard supports some of the weight of the overlying sediments. When groundwater is removed from aquifers by excessive pumping, pore pressures in the aquifer drop and compression of the aquifer may occur. This compression may be partially recoverable if pressures rebound, but much of it is not. When the aquifer gets compressed, it may cause land subsidence, a drop in the ground surface. The city of New Orleans, Louisiana is actually below sea level today, and its subsidence is partly caused by removal of groundwater from the various aquifer/aquitard systems beneath it. In the first half of the 20th century, the San Joaquin Valley experienced significant subsidence, in some places up to 8.5 metres (28 feet) due to groundwater removal. Cities on river deltas, including Venice in Italy, and Bangkok in Thailand, have experienced surface subsidence; Mexico City, built on a former lake bed, has experienced rates of subsidence of up to 40 cm (1'3") per year.

Seawater Intrusion

In general, in very humid or undeveloped regions, the shape of the water table mimics the slope of the surface. The recharge zone of an aquifer near the seacoast is likely to be inland, often at considerable distance. In these coastal areas, a lowered water table may induce sea water to reverse the flow toward the land. Sea water moving inland is called a saltwater intrusion. In alternative fashion, salt from mineral beds may leach into the groundwater of its own accord.

Pollution

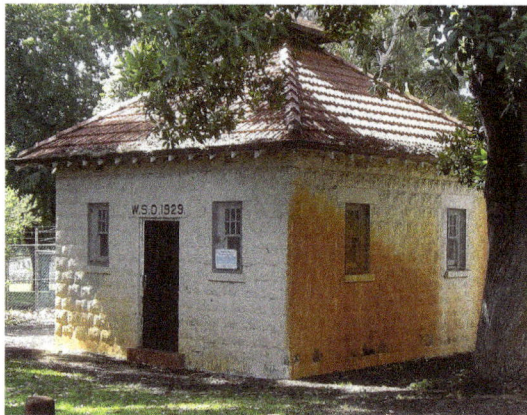

Iron oxide staining caused by reticulation from an unconfined aquifer in karst topography. Perth, Western Australia.

Polluted groundwater is less visible, but more difficult to clean up, than pollution in rivers and lakes. Groundwater pollution most often results from improper disposal of wastes on land. Major sources include industrial and household chemicals and garbage landfills, industrial waste lagoons, tailings and process wastewater from mines, oil field brine pits, leaking underground oil storage tanks and pipelines, sewage sludge and septic systems. Polluted groundwater is mapped by sampling soils and groundwater near suspected or known sources of pollution, to determine the extent of the pollution, and to aid in the design of groundwater remediation systems. Preventing groundwater pollution near potential sources such as landfills requires lining the bottom of a landfill with watertight materials, collecting any leachate with drains, and keeping rainwater off any potential contaminants, along with regular monitoring of nearby groundwater to verify that contaminants have not leaked into the groundwater.

Groundwater pollution, from pollutants released to the ground that can work their way down into groundwater, can create a contaminant plume within an aquifer. Pollution can occur from landfills, naturally occurring arsenic, on-site sanitation systems or other point sources, such as petrol stations or leaking sewers.

Movement of water and dispersion within the aquifer spreads the pollutant over a wider area, its advancing boundary often called a plume edge, which can then intersect with groundwater wells or daylight into surface water such as seeps and springs, making the water supplies unsafe for humans and wildlife. Different mechanism have influence on the transport of pollutants, e.g. diffusion, adsorption, precipitation, decay, in the groundwater. The interaction of groundwater contamination with surface waters is analyzed by use of hydrology transport models.

The danger of pollution of municipal supplies is minimized by locating wells in areas of deep groundwater and impermeable soils, and careful testing and monitoring of the aquifer and nearby potential pollution sources.

Arsenic and Fluoride

Around one-third of the world's population drinks water from groundwater resources. Of this, about 10 percent, approximately 300 million people, obtains water from groundwater resources that are heavily polluted with arsenic or fluoride. These trace elements derive mainly from natural sources by leaching from rock and sediments.

New method of Identifying Substances that are Hazardous to Health

In 2008, the Swiss Aquatic Research Institute, Eawag, presented a new method by which hazard maps could be produced for geogenic toxic substances in groundwater. This provides an efficient way of determining which wells should be tested.

In 2016, the research group made its knowledge freely available on the Groundwater Assessment Platform GAP. This offers specialists worldwide the possibility of uploading their own measurement data, visually displaying them and producing risk maps for areas of their choice. GAP also serves as a knowledge-sharing forum for enabling further development of methods for removing toxic substances from water.

Regulations

United States

In the United States, laws regarding ownership and use of groundwater are generally state laws; however, regulation of groundwater to minimize pollution of groundwater is by both states and the federal-level Environmental Protection Agency. Ownership and use rights to groundwater typically follow one of three main systems:

- The Rule of Capture provides each landowner the ability to capture as much groundwater as they can put to a beneficial use, but they are not guaranteed any set amount of water. As a result, well-owners are not liable to other land-owners for taking water from beneath their land. State laws or regulations will often define "beneficial use", and sometimes place other limits, such as disallowing groundwater extraction which causes subsidence on neighboring property.

- Limited private ownership rights similar to riparian rights in a surface stream. The amount of groundwater right is based on the size of the surface area where each landowner gets a corresponding amount of the available water. Once adjudicated, the maximum amount of the water right is set, but the right can be decreased if the total amount of available water decreases as is likely during a drought. Landowners may sue others for encroaching upon their groundwater rights, and water pumped for use on the overlying land takes preference over water pumped for use off the land.

- In November 2006, the Environmental Protection Agency published the groundwater Rule in the United States Federal Register. The EPA was worried that the groundwater system would be vulnerable to contamination from fecal matter. The point of the rule was to keep microbial pathogens out of public water sources. The 2006 groundwater Rule was an amendment of the 1996 Safe Drinking Water Act.

Other rules in the United States include:

- Reasonable Use Rule (American Rule): This rule does not guarantee the land-owner a set amount of water, but allows unlimited extraction as long as the result does not unreasonably damage other wells or the aquifer system. Usually this rule gives great weight to historical uses and prevents new uses that interfere with the prior use.

- Groundwater scrutiny upon real estate property transactions in the US: In the US, upon commercial real estate property transactions both groundwater and soil are the subjects of scrutiny. For brownfields sites (formerly contaminated sites that have been remediated), Phase I Environmental Site Assessments are typically prepared, to investigate and disclose potential pollution issues. In the San Fernando Valley of California, real estate contracts for property transfer below the Santa Susana Field Laboratory (SSFL) and eastward have clauses releasing the seller from liability for groundwater contamination consequences from existing or future pollution of the Valley Aquifer.

References

- Glaser, P. H. (1992). Raised bogs in eastern North America: regional controls for species richness and floristic assemblages. Journal of Ecology, 80, 535–54

- W. S. B. Paterson (1994). Physics of Glaciers (3rd ed.). Pergamon Press. ISBN 0-08-013972-8. OCLC 26188. A comprehensive reference on the physical principles underlying formation and behavior

- De Róiste, Daithí. "Bord na Móna announces biggest change of land use in modern Irish history". Bord na Móna. Bord na Móna. Retrieved 6 October 2015

- Meybeck, Michel (1993). "Riverine transport of atmospheric carbon: Sources, global typology and budget". Water, Air, & Soil Pollution. 70 (1–4): 443–463. doi:10.1007/BF01105015

- Damman, A. W. H. (1986). Hydrology, development, and biogeochemistry of ombrogenous bogs with special reference to nutrient relocation in a western Newfoundland bog. Canadian Journal of Botany, 64, 384–94

- Keddy, P.A. 2010. Wetland Ecology: Principles and Conservation, (2nd edition). Cambridge University Press, ISBN 978-0521739672

- "Information Sheet on Ramsar Wetlands (RIS)". Ramsar Convention on Wetlands. Archived from the original on March 4, 2009. Retrieved 2 March 2013

- Zektser, S.; LoaIciga, H. A.; Wolf, J. T. (2004). "Environmental impacts of groundwater overdraft: selected case studies in the southwestern United States". Environmental Geology. 47 (3): 396–404. doi:10.1007/s00254-004-1164-3

- Witham, F. and Llewellin, E.W., 2006. Stability of lava lakes. Journal of Volcanology and Geothermal Research, 158(3), pp. 321-332

- Thomas V. Cech (2009). Principles of Water Resources: History, Development, Management, and Policy. John Wiley & Sons. p. 83. ISBN 978-0-470-13631-7

- Chu, Jennifer (July 2012). "River networks on Titan point to a puzzling geologic history". MIT Research. Retrieved 24 July 2012

- Tosi, Luigi; Teatini, Pietro; Strozzi, Tazio; Da Lio, Cristina (2014). "Relative Land Subsidence of the Venice Coastland, Italy": 171–73. doi:10.1007/978-3-319-08660-6_32

- Sophocleous, Marios (2002). "Interactions between groundwater and surface water: the state of the science". Hydrogeology Journal. 10: 52–67. Bibcode:2002HydJ...10...52S. doi:10.1007/s10040-001-0170-8

- Poehls, D.J. and Smith, G.J. eds., 2009. Encyclopedic dictionary of hydrogeology. Academic Press. p. 517. ISBN 978-0-12-558690-0

- "What is hydrology and what do hydrologists do?". The USGS Water Science School. United States Geological Survey. 23 May 2013. Retrieved 21 Jan 2014

- Winkel, L.; Berg, M.; Amini, M.; Hug, S.J.; Johnson, C.A. Predicting groundwater arsenic contamination in Southeast Asia from surface parameters. Nature Geoscience, 1, 536–42 (2008). doi:10.1038/ngeo254

3

Classification of Freshwater Plants

Freshwater plants are classified into Acorus, Alisma, Cabomba and Myriophyllum. Acorus are plants that are found in wetlands and depend on Aerenchyma to supply oxygen to the roots whereas Cabomba are aquatic plants. The major categories of freshwater plants are dealt with great details in the chapter.

Acorus

Acorus is a genus of monocot flowering plants. This genus was once placed within the family Araceae (aroids), but more recent classifications place it in its own family Acoraceae and order Acorales, of which it is the sole genus of the oldest surviving line of monocots. Some older studies indicated that it was placed in a lineage (the order Alismatales), that also includes aroids (Araceae), Tofieldiaceae, and several families of aquatic monocots (e.g., Alismataceae, Posidoniaceae). However, modern phylogenetic studies demonstrate that *Acorus* is sister to all other monocots. Common names include calamus and sweet flag.

The genus is native to North America and northern and eastern Asia, and naturalised in southern Asia and Europe from ancient cultivation. The known wild populations are diploid except for some tetraploids in eastern Asia, while the cultivated plants are sterile triploids, probably of hybrid origin between the diploid and tetraploid forms.

Characteristics

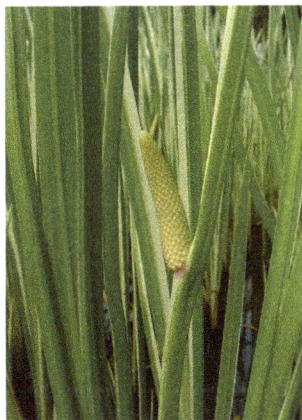

Habit of *Acorus calamus*.

The inconspicuous flowers are arranged on a lateral spadix (a thickened, fleshy axis). Unlike aroids, there is no spathe (large bract, enclosing the spadix). The spadix is 4–10 cm long and is enclosed by the foliage. The bract can be ten times longer than the spadix. The leaves are linear with entire margin.

Taxonomy

Although the family Acoraceae was originally described in 1820, since then *Acorus* has traditionally been included in Araceae in most classification systems, as in the Cronquist system. The family has recently been resurrected as molecular systematic studies have shown that *Acorus* is not closely related to Araceae or any other monocot family, leading plant systematists to place the genus and family in its own order. This placement currently lacks support from traditional plant morphology studies, and some taxonomists still place it as a subfamily of Araceae, in the order Alismatales. The APG III system recognizes order Acorales, distinct from the Alismatales, and as the sister group to all other monocots. This relationship is confirmed by more recent phylogenetic studies. Treatment in the APG IV system is unchanged from APG III.

Species

In older literature and on many websites, there is still much confusion, with the name *Acorus calamus* equally but wrongfully applied to *Acorus americanus* (formerly *Acorus calamus* var. *americanus*).

As of July 2014, the Kew Checklist accepts only 2 species, one of which has three accepted varieties:

- *Acorus calamus* L. – Common sweet flag; sterile triploid ($3n = 36$); probably of cultivated origin. It is native to Europe, temperate India and the Himalayas and southern Asia, widely cultivated and naturalised elsewhere.

  o *Acorus calamus* var. *americanus* Raf. - Canada, northern United States, Buryatiya region of Russia

  o *Acorus calamus* var. *angustatus* Besser - Siberia, China, Russian Far East, Japan, Korea, Mongolia, Himalayas, Indian Subcontinent, Indochina, Philippines, Indonesia

  o *Acorus calamus* var. *calamus* - Siberia, Russian Far east, Mongolia, Manchuria, Korea, Himalayas; naturalized in Europe, North America, Java and New Guinea

- *Acorus gramineus* Sol. ex Aiton – Japanese sweet flag or grassy-leaved sweet flag; fertile diploid ($2n = 18$); - China, Himalayas, Japan, Korea, Indochina, Philippines, Primorye

Acorus from Europe, China and Japan have been planted in the United States.

Etymology

The name 'acorus' is derived from the Greek word 'acoron', a name used by Dioscorides, which in turn was derived from 'coreon', meaning 'pupil', because it was used in herbal medicine as a treatment for inflammation of the eye.

Distribution and habitat

These plants are found in wetlands, particularly marshes, where they spread by means of thick rhizomes. Like many other marsh plants, they depend upon aerenchyma to transport oxygen to the rooting zone. They frequently occur on shorelines and flood-plains where water levels fluctuate seasonally.

Ecology

The native North American species appears in many ecological studies. Compared to other species of wetland plants, they have relatively high competitive ability. Although many marsh plants accumulate large banks of buried seeds, seed banks of *Acorus* may not accumulate in some wetlands owing to low seed production. The seeds appear to be adapted to germinate in clearings; after a period of cold storage, the seeds will germinate after seven days of light with fluctuating temperature, and somewhat longer under constant temperature. A comparative study of its life history traits classified it as a "tussock interstitial", that is, a species that has a dense growth form and tends to occupy gaps in marsh vegetation, not unlike *Iris versicolor*.

Toxicity

Products derived from *Acorus calamus* were banned in 1968 as food additives by the United States Food and Drug Administration. The questionable chemical derived from the plant was β-asarone. Confusion exists whether all strains of *A. calamus* contain this substance.

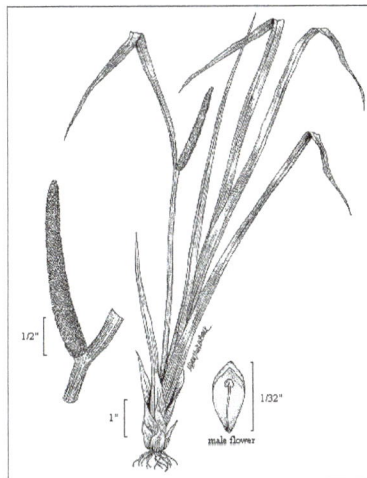

Sweet Flag (2006 drawing by USGS Northern Prairie Wildlife Research Center)

Four varieties of *A. calamus* strains exist in nature: diploid, triploid, tetraploid and hexaploid. Diploids do not produce the carcinogenic β-asarone. Diploids are known to grow naturally in Eastern Asia (Mongolia and C Siberia) and North America. The triploid cytotype probably originated in the Himalayan region, as a hybrid between the diploid and tetraploid cytotypes. The North American Calamus is known as *Acorus calamus* var. *americanus* or more recently as simply *Acorus americanus*. Like the diploid strains of *A. calamus* in parts of the Himalayas, Mongolia, and C Siberia, the North American diploid strain does not contain the carcinogenic β-asarone. Research has consistently demonstrated that "β-asarone was not detectable in the North American spontaneous diploid Acorus [Calamus var. Americanus]".

Uses

The parallel-veined leaves of some species contain ethereal oils that give a sweet scent when dried. Fine-cut leaves used to be strewn across the floor in the Middle Ages, both for the scent, and for presumed efficacy against pests.

Acorus Gramineus

Acorus gramineus, commonly known as grass-leaf sweet flag, dwarf sedge, Japanese rush, and Japanese sweet flag, is a botanical species belonging to the genus Acorus, native to Japan, Korea in eastern Asia. The plant usually grows in wetlands and shallow water.

Description

This shrubby plant's long, narrow, slightly curved leaves may grow to 30 cm (12 inches) in height. It can grow fully or partially submerged, or in very moist soil, but it will usually only flower when at least partially submerged.

Var. *pusillus* has slightly shorter, more rigid glossy green leaves, while var. *variegatus* has longer leaves streaked with yellow.

Cultivation and Uses

Acorus gramineus spreads aggressively by rhizome, creating a nearly-seamless groundcover where conditions are favorable, and it is frequently used around the edges of ponds and water gardens, as well as submerged in freshwater aquaria. It can be propagated by dividing the fleshy underwater rhizome and planting the base in shallow water.

In Japan during the Heian period, leaves of the plant were gathered for the Sweet Flag Festival on the fifth day of the fifth month. Sweet flag and wormwood were spread on the roofs of houses for decoration and to ward off evil spirits. Special herbal balls made of sweet flag were also fashioned for the occasion.

Acorus Calamus

Acorus calamus (also called sweet flag or calamus, among many common names) is a tall perennial wetland monocot of the Acoraceae family, in the genus *Acorus*.

Description

Sweet Sedge is a perennial herb, 30 to 100 cm tall. In habit it resembles the Iris. It consists of tufts of basal leaves that rise from a spreading rhizome. The leaves are erect yellowish-green, radical, with pink sheathing at their bases, sword-shaped, flat and narrow, tapering into a long, acute point, and have parallel veins. The leaves have smooth edges, which can be wavy or crimped. The sweet sedge can easily be distinguished from the Iris and other similar plants, by the unusual crimped edges of the leaves, the fragant odour it emits when crushed, and the unusual flower spadix.

The solid, triangular flower-stems rise from the axils of the outer leaves. A semi-erect spadix emerges from one side of the flower stem. The spadix is solid, cylindrical, tapers at each end, and is 5 to 10 cm in length. A covering spathe, as is usual with Acoraceae, is absent. The spadix is densely crowded with tiny greenish-yellow flowers arranged in diamond-shaped pattern. Each flower contains six petals and stamens enclosed in a perianth with six divisions and surrounding a three-celled, oblong ovary with a sessile stigma. The flowers are sweetly fragrant. In Europe it flowers for about a month in late spring or early summer, but usually does not bear fruit there. Only plants that grow in water bear flowers. The fruit is a berry filled with mucus, which when ripe falls into the water and thus disperses. Even in Asia it fruits sparingly and propagates itself mainly by spreading its rhizome, forming colonies.

The branched, cylindrical, knobby rhizome is about a finger thick and has numerous coarse fibrous roots below it. The exterior is brown and the interior white.

Range and Habitat

Sweet Flag is native to India, central Asia, southern Russia and Siberia, and perhaps Eastern Europe. It also grows in China and Japan. It was introduced into Western Europe and North America for medicinal purposes. Habitats include edges of small lakes, ponds and rivers, marshes, swamps, and wetlands.

Names

In addition to "sweet flag" and "calamus" other common names include beewort, bitter pepper root, calamus root, flag root, gladdon, myrtle flag, myrtle grass, myrtle root, myrtle sedge, pine root, rat root, sea sedge, sweet cane, sweet cinnamon, sweet grass, sweet myrtle, sweet root, sweet rush, and sweet sedge.

Etymology

The generic name is the Latin word *acorus*, which is derived from the Greek (áchórou) of Dioscorides (note different versions of the text have different spellings). The word itself is thought to have been derived from the word κόρη (kóri), which means pupil (of an eye), because of the juice from the root of the plant being used as a remedy in diseases of the eye ('darkening of the pupil').

The specific name *calamus* (meaning "cane") is derived from Greek (kálamos, meaning "reed"), which is cognate to Latin *culmus* (meaning "stalk") and Old English *healm* (meaning "straw"), and derived from Proto-Indo European **kole-mo-* (thought to mean "grass" or "reed").

The name *sweet flag* refers to its sweet scent and its similarity to *Iris* species, which are commonly known as flags in English since the late fourteenth century.

Botanical Information

Currently the taxonomic position of these forms is contested. The comprehensive taxonomic analysis in the Kew World Checklist of Selected Plant Families from 2002 considers all three forms to be distinct varieties of a single species. Sue A. Thompson in her 1995 Ph.D. dissertation and in her 2000 entry in the Flora of North America considers the diploid form to be a distinct species. Thompson only analyses North American forms of the diploid variety in her treatment, and does not analyse the morphology of Asian forms of the diploid variety. Also, in older USA literature the name *Acorus americanus* may be used indiscriminately for all forms of *Acorus calamus* occurring in North America, irrespective of cytological diversity (i.e. both the diploid and triploid forms). The recent treatment in the Flora of China from 2010, which is followed in the Tropicos database system, considers all varieties to be synonyms of a single taxonomically undifferentiated species, pointing to morphological overlap in the characteristics singled out by Thompson.

According to Thompson the primary morphological distinction between the triploid and the North American forms of the diploid is made by the number of prominent leaf veins, the diploid having a single prominent midvein and on both sides of this equally raised secondary veins, the triploid having a single prominent midvein with the secondary veins barely distinct. Thompson notes a number of other details which she claims can be used to tell the different forms apart in North America, such as flower length, average maximum leaf length, relative length of the sympodial leaf with respect to the vegetative leaves, the average length of the spadix during flowering, and tendency of the leaf margin to undulate in the triploid. She notes that many of these characteristics overlap, but that in general the triploid is somewhat larger and more robust on average than most North American forms of the diploid. According to Heng Li, Guanghua Zhu and Josef Bogner in the Flora of China there is clear overlap in these characteristics and the different cytotypes are impossible to distinguish morphologically.

Triploid plants are infertile and show an abortive ovary with a shrivelled appearance. This form will never form fruit (let alone seeds) and can only spread asexually.

The tetraploid variety is usually known as *Acorus calamus var. angustatus* Besser. A number of synonyms are known, but a number are contested as to which variety they belong. It is morphologically diverse, with some forms having very broad and some narrow leaves. It is furthermore also cytotypically diverse, with an array of different karyotypes.

Uses

A. calamus has been an item of trade in many cultures for thousands of years. It has been used medicinally for a wide variety of ailments, and its aroma makes calamus essential oil valued in the perfume industry. The essence from the rhizome is used as a flavor for pipe tobacco. When eaten in crystallized form, it is called "German ginger". In Europe *Acorus calamus* was often added to wine, and the root is also one of the possible ingredients of absinthe. It is also used in bitters. In Lithuania *Ajeras* (Sweet flag) is added to home baked black bread.

History

The Bible mentions its use in the holy anointing oil (Exodus 30: 23). Although probably not native to Egypt, this plant was already mentioned in the Chester Beatty papyrus VI dating to approximately 1300 BC. The ancient Egyptians rarely mentioned the plant in medicinal contexts (the aforementioned papyrus mentioned using it in conjunction with several ingredients as a bandage used to soothe an ailment of the stomach), but it was certainly used to make perfumes.

Initially, Europeans confused the identity and medicinal uses of the *Acorus calamus* of the Romans and Greeks with their native *Iris pseudacorus*. Thus the *Herbarius zu Teutsch*, published at Mainz in 1485, describes and includes a woodcut of this iris under the name *Acorus*. This German book is one of three possible sources for the French *Le Grant Herbier*, written in 1486, 1488, 1498 or 1508, of which an English translation was published as the *Grete Herball* by Peter Treveris in 1526, all containing the false identification of the *Herbarius zu Teutsch*. William Turner, writing in 1538, describes 'acorum' as "gladon or a flag, a yelowe floure delyce".

The plant was introduced to Britain in the late 16th century. By at least 1596 true *Acorus calamus* was grown in Britain, as it is listed in *The Catalogue*, a list of plants John Gerard grew in his garden at Holborn. Gerard notes "It prospereth exceeding well in my garden, but as yet beareth neither flowers nor stalke". Gerard lists the Latin name as *Acorus verus*, but it is evident there was still doubt about its veracity: in his 1597 herbal he lists the English common name as 'bastard calamus'.

Cultural Uses

In Britain the plant was cut for use as a sweet smelling floor covering for the packed earth floors of dwellings and churches, and stacks of rushes have been used as the centrepiece of rushbearing ceremonies for many hundreds of years. It has also been used as a thatching material for English cottages.

In modern Egypt it is thought to have aphrodisiac properties.

For the Penobscot people this was a very important root. One story goes that a sickness was plaguing the people. A muskrat spirit came to a man in a dream, telling him that he (the muskrat) was a root and where to find him. The man awoke, found the root, and made a medicine which cured the people. In Penobscot homes, pieces of the dried root were strung together and hung up for preservation. Steaming it throughout the home was thought to "kill" sickness. While they were travelling, a piece of root was kept and chewed to ward off illness.

Teton-Dakota warriors chewed the root to a paste, which they rubbed on their faces. It was thought to prevent excitement and fear when facing an enemy.

The Potawatomi people powdered the dried root and placed this up the nose to cure catarrh.

On 5 May Japanese prepare a bath with hashōbu leaves (shōbu-yu) for children to promote good health and to ward-off evil. In the Japanese calendar the day is known as Ayame no sekku.

This species also has a variety of purported uses for bone and joint related issues in the state of Sikkim of Northeastern India.

Illustration from an 1885 flora

Herbal Medicine

Sweet flag has a very long history of medicinal use in Chinese and Indian herbal traditions. The leaves, stems, and roots are used in various Siddha and Ayurvedic medicines. It is widely employed in modern herbal medicine for its sedative, laxative, diuretic, and carminative properties. It is used in Ayurveda to counter the side effects of all hallucinogens. Sweet Flag, known as "Rat Root" is one of the most widely and frequently used herbal medicines amongst the Chipewyan people.

Hallucinogenic Properties

Chewing the rootstock of the plant can cause visual hallucinations, possibly because of the presence of alpha-asarone or beta-asarone.

Horticulture

This plant is sometimes used as a pond plant in horticulture. There is at least one ornamental cultivar known; it is usually called 'Variegatus', but the RHS recommends calling it 'Argenteostriatus'.

Modern Research

Acorus calamus shows neuroprotective effect against stroke and chemically induced neurodegeneration in rats. Specifically, it has protective effect against acrylamide-induced neurotoxicity.

Both roots and leaves of *A. calamus* have shown antioxidant properties.

Acorus calamus may prove to be an effective control measure against cattle tick, Rhipicephalus (Boophilus) microplus.

A recent study showed that beta-asarone isolated from *Acorus calamus* oil inhibits adipogenesis in 3T3-L1 cells and thus reduces lipid accumulation in fat cells.

Chemistry

Both triploid and tetraploid *A. calamus* contain alpha-asarone. Other phytochemicals include:

- Beta-asarone'
- eugenol

Diploids do not contain beta-asarone (β-asarone).

Cultural Symbolism

The calamus has long been a symbol of love. The name is associated with a Greek myth:

Kalamos, son of the river-god Maeander, who loved the youth Karpos, of Zephyrus (the West Wind) and Chloris (Spring). When Karpos drowned in a swimming race, Kalamos also drowned and was transformed into a reed, whose rustling in the wind was interpreted as a sigh of lamentation.

The plant was a favorite of Henry David Thoreau (who called it "sweet flag"), and also of Walt Whitman, who added a section called the "Calamus" poems, to the third edition of *Leaves of Grass* (1860). In the poems the calamus is used as a symbol of love, lust, and affection.

The root of the calamus is cut into disc-shaped beads, and made into bracelets, which are typically worn by newborns for the first few months. A vasambu bracelet is a symbol of a newborn baby in Tamil culture.

Safety and Regulations

A. calamus and products derived from *A. calamus* (such as its oil) were banned from use as human food or as a food additive in 1968 by the United States Food and Drug Administration. The FDA ban was the result of lab studies that involved supplementing the diets of lab animals over a prolonged period of time with massive doses of isolated chemicals (β-asarone) from the Indian Jammu strain of calamus. The animals developed tumors, and the plant was labeled procarcinogenic. Wichtl says "It is not clear whether the observed carcinogenic effects in rats are relevant to the human organism." However, most sources advise caution in ingesting strains other than the diploid strain.

In reality β-asarone is neither hepatotoxic nor directly hepatocarcinogenic. It must first undergo metabolic l'-hydroxylation in the liver before achieving toxicity. Cytochrome P450 in the hepatocytes is responsible for secreting the hydrolyzing enzymes that convert β-asarone into genotoxic epoxide structure. Even with the activation of these metabolites, the carcinogenic potency is very low because of the rapid breakdown of epoxide residues with hydrolase which leaves these compounds inert. Additionally, the major metabolite of β-asarone is 2,4,5-trimethoxycinnamic acid, a derivative which is not a carcinogen.

Alisma

Alisma is a genus of flowering plants in the family Alismataceae, members of which are commonly known as water-plantains. The genus consists of aquatic plants with leaves either floating or submerged, found in a variety of still water habitats around the world (nearly worldwide). The flowers are hermaphrodite, and are arranged in panicles, racemes, or umbels. *Alisma* flowers have six stamens, numerous free carpels in a single whorl, each with 1 ovule, and subventral styles. The fruit is an achene with a short beak.

The nineteenth century British art and social critic John Ruskin believed that the particular curve of the leaf-ribs of *Alisma* represented a model of 'divine proportion' and helped shape his theory of Gothic architecture.

Copóg Phádraig ("*leaf of Patrick*") is the Irish name for the water-plantain. It is reputed to ward off fairies.

Water plantains are perennial plants. These herbs are usually emergent plants 0.1 – 1 m high. They have broad leaves that can be either tapered or rounded at the base. When submerged, the plant produces ribbon-like leaves. Inflorescences are highly branched. They produce whorls of perfect flowers either white or pinkish. The fruits are flat-sided nutlets 2.5 – 3 mm in length. These herbs usually flower in late May to early September, but thiscan vary with conditions.

Water-plantains are wetland plants and found in saturated soils and shallow water as well as marshes, wooded swamps, shrub swamps and flooded farmland. When introduced to an area, water plantain can rapidly reproduce.

Species and Subspecies

The following taxa are recognized as of May 2014:

- *Alisma* × *bjoerkqvistii* Tzvelev - Russia

- *Alisma canaliculatum* A.Braun & C.D.Bouché - eastern Asia

- *Alisma gramineum* Lej. - Europe, Asia, North Africa; minor naturalization in North America

- *Alisma* × *juzepczukii* Tzvelev - Russia

- *Alisma lanceolatum* Withering - Europe, Asia, North Africa; naturalized in Australia, New Zealand, California etc.

- *Alisma nanum* D.F.Cui - Xinjiang

- *Alisma orientale*

- *Alisma plantago-aquatica* L. - Europe, Asia, Africa

    o   *Alisma plantago-aquatica* subsp. *orientale* (Sam.) Sam.

    o   *Alisma plantago-aquatica* subsp. *plantago-aquatica*

- *Alisma* × *rhicnocarpum* Schotsman - western Europe

- *Alisma subcordatum* Raf. eastern North America

- *Alisma triviale* Pursh - North America

- *Alisma wahlenbergii* (Holmb.) Juz. Baltic Sea region

Alisma Gramineum

Alisma gramineum is a small aquatic plant in the water-plantain family. It has several common names including narrowleaf water-plantain, ribbonleaf water-plantain or ribbon-leaved water-plantain, and grass-leaved water-plantain. It grows in mud or submerged in shallow fresh or brackish water in marshy areas.

Description

The leaves and tiny purple-tinted white flowers may be submersed or not. When the flowers grow underwater they are cleistogamous, meaning they stay closed and self-pollinate. When the flowers grow above water they open. The leaves above the surface are stiff and wide, but submerged leaves are ribbon-like. The fruit is a ring of dry nutlets. Reproduction is by seed or from division of the corm.

Distribution

Alisma gramineum is widespread across temperate and subarctic portions of Asia and Europe and North Africa from France and Libya to China and Yakutsk. It is reported from much of Canada from British Columbia to Quebec, as well as most of the western United States plus New York, Vermont and Virginia. This is an endangered and protected species in the United Kingdom.

Alisma Lanceolatum

Alisma lanceolatum is a species of aquatic plant in the water plantain family known by the common names lanceleaf water plantain and narrow-leaved water plantain. It is widespread across Europe, North Africa and temperate Asia. It is naturalized in Australia, New Zealand, Oregon, California and British Columbia. It is considered a noxious weed in some places.

This species is a weed of rice fields in many areas, including New South Wales and California.

In England and Wales it is occasionally locally found, in Ireland it is rare, and Scotland it is very rare.

It is found in mud and in fresh waters.

Description

This is a perennial herb growing from a caudex in the water or mud. It produces lance-shaped leaves 12 to 20 centimeters long and 4 wide on long petioles; leaves which remain submerged in water are smaller and less prominently veined. The inflorescence is mostly erect and up to half a meter tall.

Part of an inflorescence and single blossom

It is a wide array of small pink-petalled flowers, which open in the morning, from June until August. The fruit is a tiny achene up to 2 or 3 millimeters long clustered into an aggregate fruit of about 20 units.

Similar Species

The water plantain *Alisma plantago-aquatica* has acute leaf tips not tapering to a stalk.

Cabomba

Cabomba is an aquatic plant genus, one of two belonging to the family Cabombaceae. It has divided submerged leaves in the shape of a fan (hence the vernacular name fan-wort) and is much favoured by aquarists as an ornamental and oxygenating plant for fish tanks. Use in the aquarium trade has led to some species being introduced to other parts of the world, such as Australia, where they have become weeds.

Species

- *Cabomba aquatica* Aubl. (fanwort)

- *Cabomba caroliniana* A. Gray (green cabomba)

- *Cabomba furcata* Schult. & Schult.f. (red cabomba)

- *Cabomba haynesii* Wiersema

- *Cabomba palaeformis* Fassett

Cabomba as an Aquarium Plant

Cabomba is frequently planted in aquaria, as an attractive-leaved water plant that is fast-growing (up to one inch per day). Green cabomba (*C. caroliniana*) is the most common, and the easiest aquarium subject. By contrast, red cabomba (*C. furcata*) is considered to be one of the hardest plants to care for in the aquarium.

Flowers and Reproduction

The perianth of Cabomba is either trimerous (having members in each whorl in groups of three) or bimerous (in groups of two) with white, oval-shaped petals, and is usually about 2.0 cm across when fully developed. The petals are unlike the sepals in that the former have two yellow ear-shaped nectaries at the base. Petals may also have purplish edges. Flowers are protogynous, having primarily female sexual structures on the first day of appearance and then switching to male on the second and subsequent days. Flowers emerge and are designed to be pollinated above the waterline. Principal pollinators are flies and other small flying insects.

Cabomba Caroliniana

Cabomba caroliniana is an aquatic perennial herbaceous plant native to North and South America. It is a weed of national significance in Australia and on the list of invasive alien species of union concern in the EU.

Common Names

Cabomba caroliniana is commonly called Carolina fanwort, Carolina water shield, green cabomba, fanwort, fish grass, and Washington grass.

Distribution

It is native to southeastern South America (southern Brazil, Paraguay, Uruguay, and northeastern Argentina), and the East and West Coasts of the United States. It is eaten as a vegetable in some areas.

Ecological Aspects

This species grows rooted in the mud of stagnant to slow-flowing water, including streams, smaller rivers, lakes, ponds, sloughs, and ditches. In some states in the United States, it is now regarded as a weed. Fanwort stems become brittle in late summer, which causes the plant to break apart, facilitating its distribution and invasion of new water bodies. It produces by seed, but vegetative reproduction seems to be its main vehicle for spreading to new waters. Growth of 50 mm (2.0 in) a day has been reported in Lake Macdonald in Queensland, Australia.

Large numbers of plants are sent from Florida to the rest of the U.S. for commercial use. Fanwort is also grown commercially in Asia for export to Europe and other parts of the world. Small-scale, local cultivation occurs in some areas, and aquarists are probably responsible for some introductions.

Description

Fanwort is a submerged, sometimes floating, but often rooted, freshwater perennial plant with short, fragile rhizomes. The erect shoots are upturned extensions of the horizontal rhizomes. The shoots are grass-green to olive-green or sometimes reddish brown. The leaves are of two types: submerged and floating. The submerged leaves are finely divided and arranged in pairs on the stem. The floating leaves, when present, are linear and inconspicuous, with an alternate arrangement. They are less than $\frac{1}{2}$ in (13 mm) long and narrow (less than $\frac{1}{4}$ in or 6.4 mm). The leaf blade attaches to the centre, where a slight constriction is seen. The flowers are white and small (less than $\frac{1}{2}$ in (13 mm) in diameter), and are on stalks which arise from the tips of the stems.

Cabomba Furcata

Cabomba furcata is a species of aquatic plant in the water shield family known by the common names red cabomba and forked fanwort. It is native to South America and as far north as Cuba and the tip of Florida. It reaches a maximum height between 30 and 80 cm and is up to 8 cm wide. It bears purple flowers.

This is used as an aquarium plant. Carbon dioxide addition is usually necessary, mostly because this plant requires high light and regular fertilization for optimal growth.

References

- Leck, Mary Allessio; Simpson, Robert L. (1995). "Ten-year seed bank and vegetation dynamics of a tidal freshwater marsh". American Journal of Botany. 82: 1547–1557. doi:10.2307/2446183

- Simon & Schuster's Guide to Freshwater and Marine Aquarium Fishes. New York, New York, United States: Simon & Schuster, Inc. 1976. ISBN 0-671-22809-9

- Lansdown, R.V. (2014). "Acorus calamus". IUCN Red List of Threatened Species. Version 2014.2. International Union for Conservation of Nature. Retrieved 28 August 2014

- Gaudet, C.L.; Keddy, P.A. (1988). "Predicting competitive ability from plant traits: a comparative approach". Nature. 334: 242–243. doi:10.1038/334242a0

- Shipley, B.; Keddy, P.A.; Moore, D.R.J.; Lemky, K. (1990). "Regeneration and establishment strategies of emergent macrophytes". Journal of Ecology. 77: 1093–1110. doi:10.2307/2260825. Appendix 3

- Sei Shōnagon (2006). The Pillow Book. Translated by Meredith McKinney. London, England: Penguin Books, Ltd. pp. 41–42, 282. ISBN 0-140-44806-3

- "BSBI List 2007". Botanical Society of Britain and Ireland. Archived from the original (xls) on 2015-01-25. Retrieved 2014-10-17

- Van der Valk, A. G.; Davis, C. B. (1978). "The role of seed banks in the vegetation dynamics of prairie glacial marshes". Ecology. 59: 322–35. doi:10.2307/1936377

- Boutin, C.; Keddy, P. A. (1993). "A functional classification of wetland plants". Journal of Vegetation Science. 4: 591–600. doi:10.2307/3236124

- Hirsch, Pamela; Gladstar, Rosemary (2000). Planting the future: saving our medicinal herbs. Rochester, Vt: Healing Arts Press. p. 85. ISBN 0-89281-894-8

- O'Neill, Alexander; et al. (2017-03-29). "Integrating ethnobiological knowledge into biodiversity conservation in the Eastern Himalayas". Journal of Ethnobiology and Ethnomedicine. 13 (21). doi:10.1186/s13002-017-0148-9. Retrieved 2017-05-11

- Avadhani, Mythili; et al. (2013). "The Sweetness and Bitterness of Sweet Flag [Acorus calamus L.] – A Review" (PDF). Research Journal of Pharmaceutical, Biological and Chemical Sciences. 4 (2): 598. ISSN 0975-8585

- McGuffin, Michael, ed. (1997). American Herbal Products Association's Botanical Safety Handbook. Boca Raton, Florida: CRC Press. p. 135. ISBN 978-0-8493-1675-3

- Sylvan T. Runkel; Alvin F. Bull (2009) [1979]. Wildflowers of Iowa Woodlands. Iowa City, Iowa: University of Iowa Press. p. 119. Retrieved 13 December 2011

- Raina, V. K.; Srivastava, S. K.; Syamasunder, K. V.; et al. (2003). "Essential oil composition of Acorus calamus L. from the lower region of the Himalayas". Flavour and Fragrance Journal. 18 (1): 18–20. doi:10.1002/ffj.1136

4

Freshwater Animals: An Overview

Freshwater molluscs live in both flowing water as well as still water. The classes of molluscs can be divided into bivalves, clams, molluscs, etc. Freshwater fish, heron and freshwater crocodile are some of the other topics discussed in this section. This chapter is an overview of the subject matter incorporating all the major categories of freshwater animals.

Freshwater Mollusc

Bithynia tentaculata, a small freshwater gastropod in the family Bithyniidae

Freshwater molluscs are those members of the Phylum Mollusca which live in freshwater habitats, both lotic (flowing water) such as rivers, streams, canals, springs, and underground cave streams (stygobite species) and lentic (still water) such as lakes, ponds (including temporary or vernal ponds), and ditches.

The two major classes of molluscs have representatives in freshwater: the gastropods (snails) and the bivalves (freshwater mussels and clams.) It appears that the other classes within the Phylum Mollusca -the cephalopods, scaphopods, polyplacophorans, etc. - never made the transition from a fully marine environment to a freshwater environment.

A few species of freshwater molluscs are among the most notorious invasive species. In contrast, numerous others have become threatened or have become extinct in the face of anthropogenic change.

Freshwater bivalve *Alasmidonta raveneliana*

Biogeography

Typical freshwater species (such as many river mussel species in the family Unionidae) have a range which may consist of a series of adjacent river systems, a series of adjacent tributaries, or part of a single large river system. Large rivers and small tributary creeks typically share few species, and distribution patterns suggest large lowland rivers represent substantial barriers to the dispersal of species adapted to small upland streams. Endemism is common in some families, and species may be endemic to a single creek or spring. In contrast, some of the tiny pill clams have a nearly worldwide distribution (Burch, 1972)

Ecological and Anatomical Challenges

Challenges in the natural environment faced by freshwater Mollusca include floods, droughts, siltation, extreme temperature variations, predation, and the constant uni-directional flow characteristic of river habitats. Osmoregulation, or the maintenance of constant salinity within body tissue and fluids, is another challenge faced by freshwater Mollusca. Dillon (2000) indicates that they have characteristically low tissue salinities relative to other freshwater animals, and unionoid mussels have some of the lowest tissue salinities of any animal.

Freshwater Bivalves

Families of freshwater bivalves occur within the orders Unionoida and Veneroida.

Freshwater Gastropods

Ten families of prosobranchiate snails (gilled operculate snails) and five pulmonate families (lunged snails, distantly related to common landsnails) inhabit freshwater environments in many parts of the world. Some freshwater snail species serve as hosts for human and animal parasites.

Bivalvia

Bivalvia, in previous centuries referred to as the Lamellibranchiata and Pelecypoda, is a class of marine and freshwater molluscs that have laterally compressed bodies enclosed by a shell consisting of two hinged parts. Bivalves as a group have no head and they lack some usual molluscan organs like the radula and the odontophore. They include the clams, oysters, cockles, mussels, scallops, and numerous other families that live in saltwater, as well as a number of families that live in freshwater. The majority are filter feeders. The gills have evolved into ctenidia, specialised organs for feeding and breathing. Most bivalves bury themselves in sediment where they are relatively safe from predation. Others lie on the sea floor or attach themselves to rocks or other hard surfaces. Some bivalves, such as the scallops and file shells, can swim. The shipworms bore into wood, clay, or stone and live inside these substances.

Empty shell of the giant clam
(*Tridacna gigas*)

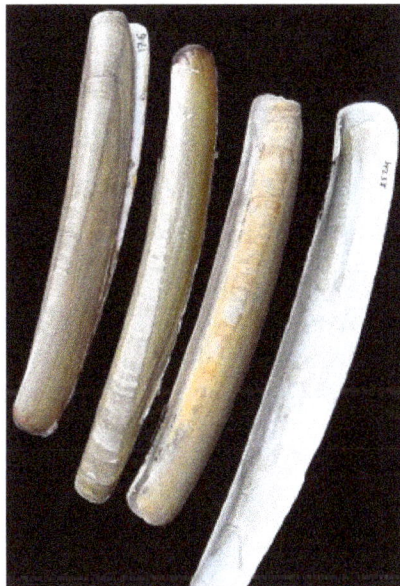

Empty shells of the sword razor
(*Ensis ensis*)

The shell of a bivalve is composed of calcium carbonate, and consists of two, usually similar, parts called valves. These are joined together along one edge (the hinge line) by a flexible ligament that, usually in conjunction with interlocking "teeth" on each of the valves, forms the hinge. This arrangement allows the shell to be opened and closed without the two halves detaching. The shell is typically bilaterally symmetrical, with the hinge lying in the sagittal plane. Adult shell sizes of bivalves vary from fractions of a millimetre to over a metre in length, but the majority of species do not exceed 10 cm (4 in).

Bivalves have long been a part of the diet of coastal and riparian human populations. Oysters were cultured in ponds by the Romans, and mariculture has more recently become an important source of bivalves for food. Modern knowledge of molluscan reproductive cycles has led to the development of hatcheries and new culture techniques. A better understanding of the potential hazards of eating raw or undercooked shellfish has led to improved storage and processing. Pearl oysters (the common name of two very different families in salt water and fresh water) are the most common source of natural pearls. The shells of bivalves are used in craftwork, and the manufacture of jewellery and buttons. Bivalves have also been used in the biocontrol of pollution.

Bivalves appear in the fossil record first in the early Cambrian more than 500 million years ago. The total number of living species is about 9,200. These species are placed within 1,260 genera and 106 families. Marine bivalves (including brackish water and estuarine species) represent about 8,000 species, combined in four subclasses and 99 families with 1,100 genera. The largest recent marine families are the Veneridae, with more than 680 species and the Tellinidae and Lucinidae, each with over 500 species. The freshwater bivalves include seven families, the largest of which are the Unionidae, with about 700 species.

Etymology

The taxonomic term Bivalvia was first used by Linnaeus in the 10th edition of his *Systema Naturae* in 1758 to refer to animals having shells composed of two valves. More recently, the class was known as Pelecypoda, meaning "axe-foot" (based on the shape of the foot of the animal when extended).

The name "bivalve" is derived from the Latin *bis*, meaning "two", and *valvae*, meaning "leaves of a door". Not all animals with shells with two hinged parts are classified under Bivalvia; other animals with paired valves include certain gastropods (small sea snails in the family Juliidae), members of the phylum Brachiopoda and the minute crustaceans known as ostracods and conchostrachans.

Anatomy

Bivalves vary greatly in overall shape. Some, such as the cockles, have shells that are nearly globular; cockles can jump by bending and straightening their foot. Others, such as the razor clams, are burrowing specialists with elongated shells and a powerful foot adapted for rapid digging. The shipworms, in the family Teredinidae have greatly elongated bodies, but their shell valves are much reduced and restricted to the anterior end of the body, where they function as scraping organs that permit the animal to dig tunnels through wood.

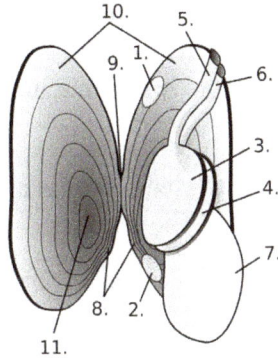

Drawing of freshwater pearl mussel (*Margaritifera margaritifera*) anatomy: 1: posterior adductor, 2: anterior adductor, 3: outer left gill demibranch, 4: inner left gill demibranch, 5: excurrent siphon, 6: incurrent siphon, 7: foot, 8: teeth, 9: hinge, 10: mantle, 11: umbo

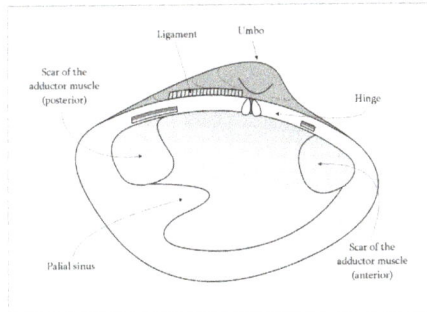

Interior of the left valve of a venerid

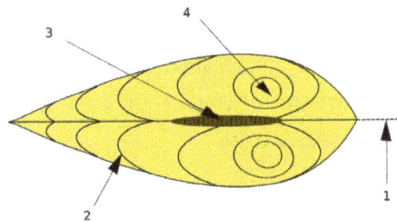

Main parts of a bivalve shell: 1: sagittal plane, 2: growth lines, 3: ligament, 4: umbo

Mantle and Shell

Near the hinge of the shell is the umbo, often a rounded, knob-like protuberance usually surrounding the beak. The umbo generally and the beak specifically represent the oldest portion of the shell, with extra material gradually being laid down along the margins on the opposite sides. The hinge point or line is the dorsal region of the shell, and the lower, curved margin is the ventral region. The anterior or front of the shell is where the byssus (when present) and foot are located, and the posterior of the shell is where the siphons are located. With the umbones/ hinge uppermost and with the anterior edge of the animal towards the viewer's left, the valve facing the viewer is the left valve and the opposing valve the right.

In all molluscs, the mantle forms a thin membrane that covers the animal's body and extends out from it in flaps or lobes. In bivalves, the mantle lobes secrete the valves, and the mantle crest secretes the whole hinge mechanism consisting of ligament, byssus threads (where present), and teeth.

Visible on the inside of most empty bivalve valves is a shiny curved line that runs more or less parallel to the outer margin of the shell and often connects the two adductor muscle scars (if the animal had two adductor muscles). This line (known as the pallial line) exists because, parallel to the opening edge of the bivalve's shell, the mantle is attached to the shell by a continuous narrow row of minute mantle retractor muscles. The function of these small muscles is to pull the loose edge of the mantle up out of harm's way when this is necessary because of minor predation attempts. In many bivalves, the mantle edges fuse at the posterior end of the shell to form two siphons, through one of which water is inhaled, and the other expelled, for respiration and suspension feeding. Often, a pocket-like space occurs into which the siphons fit when they are retracted. This is visible on the inside of the valve as an indentation on the pallial line which is known as the pallial sinus.

The shell is composed of two calcareous valves held together by a ligament. The valves are made of either calcite, as is the case in oysters, or both calcite and aragonite. Sometimes, the aragonite forms an inner, nacreous layer, as is the case in the order Pterioida. In other taxa, alternate layers of calcite and aragonite are laid down. The ligament and byssus, if calcified, are composed of aragonite. The outermost layer of the shell is the periostracum, a skin-like layer which is composed of a conchiolin. The periostracum is secreted in the groove between the outer and middle layers of the mantle, and is usually olive or brown in colour and easily abraded. The outer surface of the valves is often sculpted, with clams often having concentric striations, scallops having radial ribs and oysters a latticework of irregular markings.

The shell is added to in two ways; the valves grow larger when more material is secreted by the mantle at the margin of the shell, and the valves themselves thicken gradually throughout the animal's life as more calcareous matter is secreted by the mantle lobes. Although the (sometimes faint) concentric rings on the exterior of a valve are commonly described as "growth rings" or "growth lines", a more accurate method for determining the age of a shell is by cutting a cross section through it and examining the incremental growth bands. Use of this technique has changed views on the longevity of many bivalves. For example, the soft-shell clam (*Mya arenaria*) was thought to be short-lived, but has now been shown to have a lifespan of at least 28 years.

The two valves of the bivalve shell are held together at the hinge by a ligament composed of two keratinised proteins, tensilium and resilium. In different groups of bivalves, the ligament may be internal or external in position. The main function of the ligament (as well as joining the valves together) is to passively cause the shell to open. The shell is actively closed using the adductor muscle or muscles which are attached to the inner surface of both valves. The position of the muscles is often clearly visible on the inside of empty valves as circular or oval muscle scars. Along the hinge line of the shell are, in most cases, a number of hinge teeth which prevent the valves from moving laterally relative to one another. The arrangement of these teeth is often important in identifying bivalves.

Nervous System

The sedentary habits of the bivalves have meant that in general the nervous system is less complex than in most other molluscs. The animals have no brain; the nervous system consists of a nerve network and a series of paired ganglia. In all but the most primitive bivalves, two cerebropleural ganglia are on either side of the oesophagus. The cerebral ganglia control the sensory organs, while the pleural ganglia supply nerves to the mantle cavity. The pedal ganglia, which control the foot, are at its base, and the visceral ganglia, which can be quite large in swimming bivalves, are under the posterior adductor muscle. These ganglia are both connected to the cerebropleural ganglia by nerve fibres. Bivalves with long siphons may also have siphonal ganglia to control them.

Senses

The sensory organs of bivalves are not well developed and are largely located on the posterior mantle margins. The organs are usually mechanoreceptors or chemoreceptors, in some cases located on short tentacles. The chemoreceptor cells taste the water and are sensitive to touch. They are typically found near the siphons, but in some species, they fringe the entire mantle cavity. The osphradium is a patch of sensory cells located below the posterior adductor muscle that may serve to taste the water or measure its turbidity, but is probably not homologous with the structure of the same name found in snails and slugs. Statocysts within the organism help the bivalve to sense and correct its orientation. Each statocyst consists of a small sac lined with sensory cilia that detect the movement of a mineral mass, a statolith, under gravity. In the order Anomalodesmata, the inhalant siphon is surrounded by vibration-sensitive tentacles for detecting prey.

Many bivalves have no eyes, but a few members of the Arcoidea, Limopsoidea, Mytiloidea, Anomioidea, Ostreoidea, and Limoidea have simple eyes on the margin of the mantle. These consist of a pit of photosensory cells and a lens. Scallops have more complex eyes with a lens, a two-layered retina, and a concave mirror. All bivalves have light-sensitive cells that can detect a shadow falling over the animal.

Muscles

The main muscular system in bivalves is the posterior and anterior adductor muscles, although the anterior muscles may be reduced or even lost in some species. These strong muscles connect the two valves and contract to close the shell. They work in opposition to the ligament which tends to pull the valves apart. In sedentary or recumbent bivalves that lie on one valve, such as the oysters and scallops, the anterior adductor muscle has been lost and the posterior muscle is positioned centrally. In file shells that can swim by flapping their valves, a single, central adductor muscle occurs. These muscles are composed of two types of muscle fibres, striated muscle bundles for fast actions and smooth muscle bundles for maintaining a steady pull.

The mantle suspender muscles attach the mantle to the shell and leave an arc-shaped scar on the inside of the valve, the pallial line. The paired pedal protractor and retractor muscles operate the animal's foot. Some bivalves, such as oysters and most scallops, are unable to extend their foot and in them, these muscles are absent. Other paired muscles control the siphons and the byssus.

Circulation and Respiration

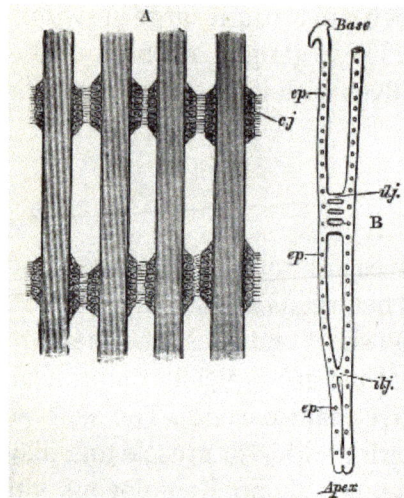

Four filaments of the gills of the blue mussel (*Mytilus edulis*) a) part of four filaments showing ciliated interfilamentar junctions (cj) b) diagram of a single filament showing the two lamellae connected at intervals by interlamellar junctions (ilj) and the position of the ciliated interfilamentar junctions (cp)

Bivalves have an open circulatory system that bathes the organs in hemolymph. The heart has three chambers: two auricles receiving blood from the gills, and a single ventricle. The ventricle is muscular and pumps hemolymph into the aorta, and then to the rest of the body. Some bivalves have a single aorta, but most also have a second, usually smaller, aorta serving the hind parts of the animal.

Oxygen is absorbed into the hemolymph in the gills which provide the primary respiratory surface. The gills hang down into the mantle cavity, the wall of which provides a secondary respiratory surface being well supplied with capillaries. In species with no gills, such as the subclass Anomalodesmata, the wall of the mantle cavity is the only organ involved in respiration. Bivalves adapted to tidal environments can survive for several hours out of water by closing their shells tightly. Some freshwater species, when exposed to the air, can gape the shell slightly and gas exchange can take place.

The hemolymph usually lacks any respiratory pigment, although members of the families Arcidae and Limidae are known to possess haemoglobin dissolved directly into the serum. In the carnivorous genus *Poromya*, the hemolymph has red amoebocytes containing a haemoglobin pigment.

Digestive System

Modes of Feeding

Most bivalves are filter feeders, using their gills to capture particulate food such as phytoplankton from the water. The protobranchs feed in a different way, scraping detritus from the seabed, and this may be the original mode of feeding used by all bivalves before the gills became adapted for filter feeding. These primitive bivalves hold on to the substratum with a pair of tentacles at the edge of the mouth, each of which has a single palp, or flap. The tentacles are covered in mucus, which traps the food, and cilia, which transport the particles back to the palps. These then sort the particles, rejecting those that are unsuitable or too large to digest, and conveying others to the mouth.

In the Filibranchia and Eulamellibranchia, water is drawn into the shell from the posterior ventral surface of the animal, passes upwards through the gills, and doubles back to be expelled just above the intake. In burrowing species, there may be two elongated, retractable siphons reaching up to the seabed, one each for the inhalant and exhalant streams of water. The gills of filter-feeding bivalves are known as ctenidia and have become highly modified to increase their ability to capture food. For example, the cilia on the gills, which originally served to remove unwanted sediment, have become adapted to capture food particles, and transport them in a steady stream of mucus to the mouth. The filaments of the gills are also much longer than those in more primitive bivalves, and are folded over to create a groove through which food can be transported. The structure of the gills varies considerably, and can serve as a useful means for classifying bivalves into groups.

A few bivalves, such as the granular poromya (*Poromya granulata*), are carnivorous, eating much larger prey than the tiny microalgae consumed by other bivalves. In these animals, the gills are relatively small, and form a perforated barrier separating the main mantle cavity from a smaller chamber through which the water is exhaled. Muscles draw water in through the inhalant siphon which is modified into a cowl-shaped organ, sucking in small crustaceans and worms at the same time. The siphon can be retracted quickly and inverted, bringing the prey within reach of the mouth. The gut is modified so that large food particles can be digested.

The unusual genus, *Entovalva*, is endosymbiotic, being found only in the oesophagus of sea cucumbers. It has mantle folds that completely surround its small valves. When the sea cucumber sucks in sediment, the bivalve allows the water to pass over its gills and extracts fine organic particles. To prevent itself from being swept away, it attaches itself with byssal threads to the host's throat. The sea cucumber is unharmed.

Digestive Tract

The digestive tract of typical bivalves consists of an oesophagus, stomach, and intestine. A number of digestive glands open into the stomach, often via a pair of diverticula; these secrete enzymes to digest food in the stomach, but also include cells that phagocytose food particles, and digest them intracellularly. In filter-feeding bivalves, an elongated rod of solidified mucus referred to as the "crystalline style" projects into the stomach from an associated sac. Cilia in the sac cause the style to rotate, winding in a stream of food-containing mucus from the mouth, and churning the stomach contents. This constant motion propels food particles into a sorting region at the rear of the stomach, which distributes smaller particles into the digestive glands, and heavier particles into the intestine. Waste material is consolidated in the rectum and voided as pellets into the exhalent water stream through an anal pore. Feeding and digestion are synchronized with diurnal and tidal cycles.

Carnivorous bivalves have a greatly reduced style, and a chitinous gizzard that helps grind up the food before digestion. In other ways, their gut is similar to that of filter-feeding bivalves.

Excretory System

Like most other molluscs, the excretory organs of bivalves are a pair of nephridia. Each of these consists of a long, looped, glandular tube, which opens into the body cavity just beneath the heart, and a bladder to store urine. The pericardial glands either line the auricles of the heart or attach to the pericardium, and serve as extra filtration organs. Metabolic waste is voided from the bladders through a pair of openings near the front of the upper part of the mantle cavity, from where it joins the stream of exhalant water.

Reproduction and Development

The sexes are usually separate in bivalves but some hermaphroditism is known. The gonads are located close to the intestines, and either open into the nephridia, or through a separate pore into the mantle cavity. The ripe gonads of male and females release sperm and eggs into the water column. Spawning may take place continually or be triggered by environmental factors such as day length, water temperature, or the presence of sperm in the water. Some species are "dribble spawners", but others release their gametes in batches or all at once. Mass spawning events sometimes take place when all the bivalves in an area synchronise their release of spawn.

Fertilization is usually external. Typically, a short stage lasts a few hours or days before the eggs hatch into trochophore larvae. These later develop into veliger larvae which settle on the seabed and undergo metamorphosis into juveniles that are sometimes (for example in the case of oysters) known as "spat". In some species, such as those in the genus *Lasaea*, females draw water containing sperm in through their inhalant siphons and fertilization takes place inside the female. These species then brood the young inside their mantle cavity, eventually releasing them into the water column as veliger larvae or as crawl-away juveniles.

Most of the bivalve larvae that hatch from eggs in the water column feed on diatoms or other phytoplankton. In temperate regions, about 25% of species are lecithotrophic, depending on nutrients stored in the yolk of the egg where the main energy source is lipids. The longer the period is before the larva first feeds, the larger the egg and yolk need to be. The reproductive cost of producing these energy-rich eggs is high and they are usually smaller in number. For example, the Baltic tellin (*Macoma balthica*) produces few, high-energy eggs. The larvae hatching out of these rely on the energy reserves and do not feed. After about four days, they become D-stage larvae, when they first develop hinged, D-shaped valves. These larvae have a relatively small dispersal potential before settling out. The common mussel (*Mytilus edulis*) produces 10 times as many eggs that hatch into larvae and soon need to feed to survive and grow. They can disperse more widely as they remain planktonic for a much longer time.

Freshwater bivalves in the order Unionoida have a different lifecycle. Sperm is drawn into a female's gills with the inhalant water and internal fertilization takes place. The eggs hatch into glochidia larvae that develop within the female's shell. Later they are released and attach themselves parasitically to the gills or fins of a fish host. After several weeks they drop off their host, undergo metamorphosis and develop into juveniles on the substrate. An advantage of this to the molluscs is that they can disperse upstream along with their temporary hosts, rather than being constantly swept downstream by the water flow.

Some of the species in the freshwater mussel family, Unionidae, commonly known as pocketbook mussels, have evolved an unusual reproductive strategy. The female's mantle protrudes from the shell and develops into an imitation small fish, complete with fish-like markings and false eyes. This decoy moves in the current and attracts the attention of real fish. Some fish see the decoy as prey, while others see a conspecific. They approach for a closer look and the mussel releases huge numbers of larvae from its gills, dousing the inquisitive fish with its tiny, parasitic young. These glochidia larvae are drawn into the fish's gills, where they attach and trigger a tissue response that forms a small cyst around each larva. The larvae then feed by breaking down and digesting the tissue of the fish within the cysts. After a few weeks they release themselves from the cysts and fall to the stream bed as juvenile molluscs. The fish are relatively unharmed.

Comparison with Brachiopods

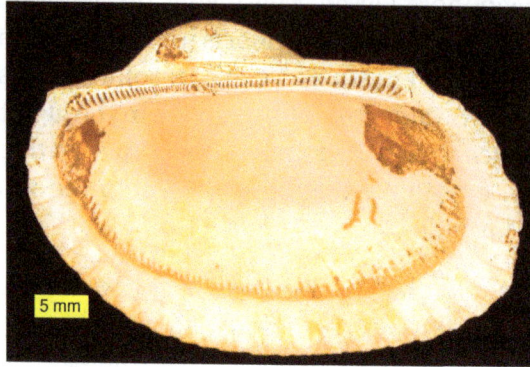

Anadara, a bivalve with taxodont dentition from the Pliocene of Cyprus

Brachiopods are shelled marine organisms that superficially resembled bivalves in that they are of similar size and have a hinged shell in two parts. However, brachiopods evolved from a very different ancestral line, and the resemblance to bivalves only arose because of a similar lifestyle. The differences between the two groups are due to their separate ancestral origins. Different initial structures have been adapted to solve the same problems, a case of convergent evolution. In modern times, brachiopods are not as common as bivalves.

A fossil Jurassic brachiopod with the lophophore support intact

Both groups have a shell consisting of two valves, but the organization of the shell is quite different in the two groups. In brachiopods, the two valves are positioned on the dorsal and ventral surfaces of the body, while in bivalves, the valves are on the left and right sides of the body, and are, in most cases, mirror images of one other. Brachiopods have a lophophore, a coiled, rigid cartilaginous internal apparatus adapted for filter feeding, a feature shared with two other major groups of marine invertebrates, the bryozoans and the phoronids. Brachiopod shells are often made of calcium phosphate as well as calcium carbonate, whereas bivalve shells are composed entirely of calcium carbonate.

Evolutionary History

The Cambrian explosion took place around 540 to 520 million years ago (Mya). In this geologically brief period, all the major animal phyla diverged and these included the first creatures with mineralized skeletons. Brachiopods and bivalves made their appearance at this time, and left their fossilized remains behind in the rocks.

Possible early bivalves include *Pojetaia* and *Fordilla*; these probably lie in the stem rather than crown group. Only five genera of supposed Cambrian "bivalves" exist, the others being *Tuarangia*, *Camya* and *Arhouriella* and potentially *Buluniella*. Bivalves have also been proposed to have evolved from the rostroconchs.

Bivalve fossils are formed when the sediment in which the shells are buried hardens into rock. Often, the impression made by the valves remains as the fossil rather than the valves. During the Early Ordovician, a great increase in the diversity of bivalve species occurred, and the dysodont, heterodont, and taxodont dentitions evolved. By the early Silurian, the gills were becoming adapted for filter feeding, and during the Devonian and Carboniferous periods, siphons first appeared, which, with the newly developed muscular foot, allowed the animals to bury themselves deep in the sediment.

By the middle of the Paleozoic, around 400 Mya, the brachiopods were among the most abundant filter feeders in the ocean, and over 12,000 fossil species are recognized. By the Permian–Triassic extinction event 250 Mya, bivalves were undergoing a huge radiation of diversity. The bivalves were hard hit by this event, but re-established themselves and thrived during the Triassic period that followed. In contrast, the brachiopods lost 95% of their species diversity. The ability of some bivalves to burrow and thus avoid predators may have been a major factor in their success. Other new adaptations within various families allowed species to occupy previously unused evolutionary niches. These included increasing relative buoyancy in soft sediments by developing spines on the shell, gaining the ability to swim, and in a few cases, adopting predatory habits.

For a long time, bivalves were thought to be better adapted to aquatic life than brachiopods were, outcompeting and relegating them to minor niches in later ages. These two taxa appeared in textbooks as an example of replacement by competition. Evidence given for this included the fact that bivalves needed less food to subsist because of their energetically efficient ligament-muscle system for opening and closing valves. All this has been broadly disproven, though; rather, the prominence of modern bivalves over brachiopods seems due to chance disparities in their response to extinction events.

Diversity of Extant Bivalves

The adult maximum size of living species of bivalve ranges from 0.52 mm (0.02 in) in *Condylonucula maya*, a nut clam, to a length of 1,532 millimetres (60.3 in) in *Kuphus polythalamia*, an elongated, burrowing shipworm. However, the species generally regarded as the largest living bivalve is the giant clam *Tridacna gigas*, which can grow to a length of 1,200 mm (47 in) and a weight of more than 200 kg (441 lb). The largest known extinct bivalve is a species of *Platyceramus* whose fossils measure up to 3,000 mm (118 in) in length.

In his 2010 treatise, *Compendium of Bivalves*, Markus Huber gives the total number of living bivalve species as about 9,200 combined in 106 families. Huber states that the number of 20,000 living species, often encountered in literature, could not be verified and presents the following table to illustrate the known diversity:

Subclass	Superfamilies	Families	Genera	Species
Heterodonta		64 (incl. 1 freshwater)	800 (16 freshwater)	5600 (270 freshwater)
	Arcticoidea	2	6	13
	Cardioidea	2	38	260
	Chamoidea	1	6	70
	Clavagelloidea	1	2	20
	Crassatelloidea	5	65	420
	Cuspidarioidea	2	20	320
	Cyamioidea	3	22	140
	Cyrenoidea	1	6 (3 freshwater)	60 (30 freshwater)
	Cyrenoidoidea	1	1	6
	Dreissenoidea	1	3 (2 freshwater)	20 (12 freshwater)
	Galeommatoidea	ca. 4	about 100	about 500
	Gastrochaenoidea	1	7	30
	Glossoidea	2	20	110
	Hemidonacoidea	1	1	6
	Hiatelloidea	1	5	25
	Limoidea	1	8	250
	Lucinoidea	2	about 85	about 500
	Mactroidea	4	46	220
	Myoidea	3	15 (1 freshwater)	130 (1 freshwater)
	Pandoroidea	7	30	250
	Pholadoidea	2	34 (1 freshwater)	200 (3 freshwater)
	Pholadomyoidea	2	3	20
	Solenoidea	2	17 (2 freshwater)	130 (4 freshwater)
	Sphaerioidea	(1 freshwater)	(5 freshwater)	(200 freshwater)
	Tellinoidea	5	110 (2 freshwater)	900 (15 freshwater)
	Thyasiroidea	1	about 12	about 100
	Ungulinoidea	1	16	100
	Veneroidea	4	104	750

Subclass	Superfamilies	Families	Genera	Species
	Verticordioidea	2	16	160
Palaeohet-erodonta		7 (incl. 6 freshwa-ter)	171 (170 freshwa-ter)	908 (900 freshwater)
	Trigonioidea	1	1	8
	Unionoidea	(6 freshwater)	(170 freshwater)	(900 freshwater)
Protobran-chia		10	49	700
	Manzanelloidea	1	2	20
	Nuculanoidea	6	32	460
	Nuculoidea	1	8	170
	Sapretoidea	1	about 5	10
	Solemyoidea	1	2	30
Pteriomor-pha		25	240 (2 freshwater)	2000 (11 freshwater)
	Anomioidea	2	9	30
	Arcoidea	7	60 (1 freshwater)	570 (6 freshwater)
	Dimyoidea	1	3	15
	Limoidea	1	8	250
	Mytiloidea	1	50 (1 freshwater)	400 (5 freshwater)
	Ostreoidea	2	23	80
	Pectinoidea	4	68	500
	Pinnoidea	1	3 (+)	50
	Plicatuloidea	1	1	20
	Pterioidea	5	9	80

Distribution

Zebra mussels encrusting a water velocity meter in Lake Michigan

The bivalves are a highly successful class of invertebrates found in aquatic habitats throughout the world. Most are infaunal and live buried in sediment on the seabed, or in the sediment in freshwater habitats. A large number of bivalve species are found in the intertidal and sublittoral zones of the oceans. A sandy sea beach may superficially appear to be devoid of life, but often a very large number of bivalves and other invertebrates are living beneath the surface of the sand. On a large beach in South Wales, careful sampling produced an estimate of 1.44 million cockles (*Cerastoderma edule*) per acre of beach.

Bivalves inhabit the tropics, as well as temperate and boreal waters. A number of species can survive and even flourish in extreme conditions. They are abundant in the Arctic, about 140 species being known from that zone. The Antarctic scallop, *Adamussium colbecki*, lives under the sea ice at the other end of the globe, where the subzero temperatures mean that growth rates are very slow. The giant mussel, *Bathymodiolus thermophilus*, and the giant white clam, *Calyptogena magnifica*, both live clustered around hydrothermal vents at abyssal depths in the Pacific Ocean. They have chemosymbiotic bacteria in their gills that oxidise hydrogen sulphide, and the molluscs absorb nutrients synthesized by these bacteria. The saddle oyster, *Enigmonia aenigmatica*, is a marine species that could be considered amphibious. It lives above the high tide mark in the tropical Indo-Pacific on the underside of mangrove leaves, on mangrove branches, and on sea walls in the splash zone.

Some freshwater bivalves have very restricted ranges. For example, the Ouachita creekshell mussel, *Villosa arkansasensis*, is known only from the streams of the Ouachita Mountains in Arkansas and Oklahoma, and like several other freshwater mussel species from the southeastern US, it is in danger of extinction. In contrast, a few species of freshwater bivalves, including the golden mussel (*Limnoperna fortunei*), are dramatically increasing their ranges. The golden mussel has spread from Southeast Asia to Argentina, where it has become an invasive species. Another well-travelled freshwater bivalve, the zebra mussel (*Dreissena polymorpha*) originated in southeastern Russia, and has been accidentally introduced to inland waterways in North America and Europe, where the species damages water installations and disrupts local ecosystems.

Behaviour

Most bivalves adopt a sedentary or even sessile lifestyle, often spending their whole lives in the area in which they first settled as juveniles. The majority of bivalves are infaunal, living under the seabed, buried in soft substrates such as sand, silt, mud, gravel, or coral fragments. Many of these live in the intertidal zone where the sediment remains damp even when the tide is out. When buried in the sediment, burrowing bivalves are protected from the pounding of waves, desiccation, and overheating during low tide, and variations in salinity caused by rainwater. They are also out of the reach of many predators. Their general strategy is to extend their siphons to the surface for feeding and respiration during high tide, but to descend to greater depths or keep their shell tightly shut when the tide goes out. They use their muscular foot to dig into the substrate. To do this, the animal relaxes its adductor muscles and opens its shell wide to anchor itself in position while it extends its foot downwards into the substrate. Then it dilates the tip of its foot, retracts the adductor muscles to close the shell, shortens its foot and draws itself downwards. This series of actions is repeated to dig deeper.

A large number of live venerid bivalves underwater with their siphons visible

Other bivalves, such as mussels, attach themselves to hard surfaces using tough byssus threads made of keratin and proteins. They are more exposed to attack by predators than the burrowing bivalves. Certain carnivorous gastropod snails such as whelks (Buccinidae) and murex snails (Muricidae) feed on bivalves by boring into their shells, although many Busyconine whelks (e.g., Busycon sinistrum, Busycon carica) are "chipping-style" predators. The dog whelk (*Nucella lamellosa*) drills a hole with its radula assisted by a shell-dissolving secretion. The dog whelk then inserts its extendible proboscis and sucks out the body contents of the victim, which is typically a blue mussel. The whelk needs a few hours to penetrate the shell and thus living in the littoral zone is an advantage to the bivalve because the gastropod can attempt to bore through the shell only when the tide is in.

Some bivalves, including the true oysters, the jewel boxes, the jingle shells, the thorny oysters and the kitten's paws, cement themselves to stones, rock or larger dead shells. In oysters the lower valve may be almost flat while the upper valve develops layer upon layer of thin horny material reinforced with calcium carbonate. Oysters sometimes occur in dense beds in the neritic zone and, like most bivalves, are filter feeders.

Bivalves filter large amounts of water to feed and breathe but they are not permanently open. They regularly shut their valves to enter a resting state, even when they are permanently submerged. In oysters, for example, their behaviour follows very strict circatidal and circadian rhythms according to the relative positions of the moon and sun. During neap tides, they exhibit much longer closing periods than during spring tides.

Although many non-sessile bivalves use their muscular foot to move around, or to dig, members of the freshwater family Sphaeriidae are exceptional in that these small clams climb about quite nimbly on weeds using their long and flexible foot. The European fingernail clam (*Sphaerium corneum*), for example, climbs around on water weeds at the edges of lakes and ponds; this enables the clam to find the best position for filter feeding.

Predators and Defence

The thick shell and rounded shape of bivalves make them awkward for potential predators to tackle. Nevertheless, a number of different creatures include them in their diet. Many species of demersal fish feed on them including the common carp (*Cyprinus carpio*), which is being used in the upper Mississippi River to try to control the invasive zebra mussel (*Dreissena polymorpha*). Birds such as the Eurasian oystercatcher (*Haematopus ostralegus*) have specially adapted beaks which can pry open their shells. The herring gull (*Larus argentatus*) sometimes drops heavy shells onto rocks in order to crack them open. Sea otters feed on a variety of bivalve species and have been observed to use stones balanced on their chests as anvils on which to crack open the shells. The Pacific walrus (*Odobenus rosmarus divergens*) is one of the main predators feeding on bivalves in Arctic waters. Shellfish have formed part of the human diet since prehistoric times, a fact evidenced by the remains of mollusc shells found in ancient middens. Examinations of these deposits in Peru has provided a means of dating long past El Niño events because of the disruption these caused to bivalve shell growth.

Invertebrate predators include crabs, starfish and octopuses. Crabs crack the shells with their pincers and starfish use their water vascular system to force the valves apart and then insert part of their stomach between the valves to digest the bivalve's body. It has been found experimentally that both crabs and starfish preferred molluscs that are attached by byssus threads to ones that are cemented to the substrate. This was probably because they could manipulate the shells and open them more easily when they could tackle them from different angles. Octopuses either pull bivalves apart by force, or they bore a hole into the shell and insert a digestive fluid before sucking out the liquified contents.

Razor shells can dig themselves into the sand with great speed to escape predation. When a Pacific razor clam (*Siliqua patula*) is laid on the surface of the beach it can bury itself completely in seven seconds and the Atlantic jackknife clam, *Ensis directus*, can do the same within fifteen seconds. Scallops and file clams can swim by opening and closing their valves rapidly; water is ejected on either side of the hinge area and they move with the flapping valves in front. Scallops have simple eyes around the margin of the mantle and can clap their valves shut to move sharply, hinge first, to escape from danger. Cockles can use their foot to move across the seabed or leap away from threats. The foot is first extended before being contracted suddenly when it acts like a spring, projecting the animal forwards.

In many bivalves that have siphons, they can be retracted back into the safety of the shell. If the siphons inadvertently get attacked by a predator, they snap off. The animal can regenerate them later, a process that starts when the cells close to the damaged site become activated and remodel the tissue back to its pre-existing form and size.

File shells such as *Limaria fragilis* can produce a noxious secretion when stressed. It has numerous tentacles which fringe its mantle and protrude some distance from the shell when it is feeding. If attacked, it sheds tentacles in a process known as autotomy. The toxin released by this is distasteful and the detached tentacles continue to writhe which may also serve to distract potential predators.

Mariculture

Oyster culture in Brittany, France

Oysters, mussels, clams, scallops and other bivalve species are grown with food materials that occur naturally in their culture environment in the sea and lagoons. One-third of the world's farmed food fish harvested in 2010 was achieved without the use of feed, through the production of bivalves and filter-feeding carps. European flat oysters (*Ostrea edulis*) were first farmed by the Romans in shallow ponds and similar techniques are still in use. Seed oysters are either raised in a hatchery or harvested from the wild. Hatchery production provides some control of the broodstock but remains problematic because disease-resistant strains of this oyster have not yet been developed. Wild spats are harvested either by broadcasting empty mussel shells on the seabed or by the use of long, small-mesh nets filled with mussel shells supported on steel frames. The oyster larvae preferentially settle out on the mussel shells. Juvenile oysters are then grown on in nursery trays and are transferred to open waters when they reach 5 to 6 millimetres (0.20 to 0.24 in) in length.

Many juveniles are further reared off the seabed in suspended rafts, on floating trays or cemented to ropes. Here they are largely free from bottom-dwelling predators such as starfish and crabs but more labour is required to tend them. They can be harvested by hand when they reach a suitable size. Other juveniles are laid directly on the seabed at the rate of 50 to 100 kilograms (110 to 220 lb) per hectare. They grow on for about two years before being harvested by dredging. Survival rates are low at about 5%.

The Pacific oyster (*Crassostrea gigas*) is cultivated by similar methods but in larger volumes and in many more regions of the world. This oyster originated in Japan where it has been cultivated for many centuries. It is an estuarine species and prefers salinities of 20 to 25 parts per thousand. Breeding programmes have produced improved stock that is available from hatcheries. A single female oyster can produce 50–80 million eggs in a batch so the selection of broodstock is of great importance. The larvae are grown on in tanks of static or moving water. They are fed high quality microalgae and diatoms and grow fast. At metamorphosis the juveniles may be allowed to settle on PVC sheets or pipes, or crushed shell. In some cases, they continue their development in "upwelling culture" in large tanks of moving water rather than being allowed to settle on the bottom. They then may be transferred to transitional, nursery beds before being moved to their final rearing quarters. Culture there takes place on the bottom, in plastic trays, in mesh bags, on rafts or on long lines, either in shallow water or in the intertidal zone. The oysters are ready for harvesting in 18 to 30 months depending on the size required.

Similar techniques are used in different parts of the world to cultivate other species including the Sydney rock oyster (*Saccostrea commercialis*), the northern quahog (*Mercenaria mercenaria*), the blue mussel (*Mytilus edulis*), the Mediterranean mussel (*Mytilus galloprovincialis*), the New Zealand green-lipped mussel (*Perna canaliculus*), the grooved carpet shell (*Ruditapes decussatus*), the Japanese carpet shell (*Venerupis philippinarum*), the pullet carpet shell (*Venerupis pullastra*) and the Yesso scallop (*Patinopecten yessoensis*).

Production of bivalve molluscs by mariculture in 2010 was 12,913,199 tons, up from 8,320,724 tons in 2000. Culture of clams, cockles and ark shells more than doubled over this time period from 2,354,730 to 4,885,179 tons. Culture of mussels over the same period grew from 1,307,243 to 1,812,371 tons, of oysters from 3,610,867 to 4,488,544 tons and of scallops from 1,047,884 to 1,727,105 tons.

Use as Food

Flat oysters (*Ostrea edulis*) from France

Bivalves have been an important source of food for humans at least since Roman times and empty shells found in middens at archaeological sites are evidence of earlier consumption. Oysters, scallops, clams, ark clams, mussels and cockles are the most commonly consumed kinds of bivalve, and are eaten cooked or raw. In 1950, the year in which the Food and Agriculture Organization (FAO) started making such information available, world trade in bivalve molluscs was 1,007,419 tons. By 2010, world trade in bivalves had risen to 14,616,172 tons, up from 10,293,607 tons a decade earlier. The figures included 5,554,348 (3,152,826) tons of clams, cockles and ark shells, 1,901,314 (1,568,417) tons of mussels, 4,592,529 (3,858,911) tons of oysters and 2,567,981 (1,713,453) tons of scallops. China increased its consumption 400-fold during the period 1970 to 1997.

It has been known for more than a century that consumption of raw or insufficiently cooked shellfish can be associated with infectious diseases. These are caused either by bacteria naturally present in the sea such as *Vibrio spp.* or by viruses and bacteria from sewage effluent that sometimes contaminates coastal waters. As filter feeders, bivalves pass large quantities of water through their gills, filtering out the organic particles, including the microbial pathogens. These are retained in the animals' tissues and become concentrated in their liver-like digestive glands. Another possible source of contamination occurs when bivalves contain marine biotoxins as a result of ingesting numerous dinoflagellates. These microalgae are not associated with sewage but occur unpredictably as algal blooms. Large areas of a sea or lake may change colour as a result of the proliferation of millions of single-cell algae, and this condition is known as a red tide.

Viral and Bacterial Infections

In 1816 in France, a physician, J. P. A. Pasquier, described an outbreak of typhoid linked to the consumption of raw oysters. The first report of this kind in the United States was in Connecticut in 1894. As sewage treatment programmes became more prevalent in the late 19th century, more outbreaks took place. This may have been because sewage was released through outlets into the sea providing more food for bivalves in estuaries and coastal habitats. A causal link between the bivalves and the illness was not easy to demonstrate because the illness might come on days or even weeks after the ingestion of the contaminated shellfish. One viral pathogen is the *Norwalk* virus. This is resistant to treatment with chlorine-containing chemicals and may be present in the marine environment even when coliform bacteria have been killed by the treatment of sewage.

In 1975 in the United States, a serious outbreak of oyster-vectored disease was caused by *Vibrio vulnificus*. Although the number of victims was low, the mortality rate was high at 50%. About 10 cases have occurred annually since then and further research needs to be done to establish the epidemiology of the infections. The cases peak in mid-summer and autumn with no cases being reported in mid winter so there may be a link between the temperature at which the oysters are held between harvesting and consumption. In 1978, an oyster-associated gastrointestinal infection affecting more than 2,000 people occurred in Australia. The causative agent was found to be the *Norwalk* virus and the epidemic caused major economic difficulties to the oyster farming industry in the country. In 1988, an outbreak of hepatitis A associated with the consumption of inadequately cooked clams (*Anadara subcrenata*) took place in the Shanghai area of China. An estimated 290,000 people were infected and there were 47 deaths.

In the United States and the European Union, since the early 1990s regulations have been in place that are designed to prevent shellfish from contaminated waters entering the food chain. This has meant that there is sometimes a shortage of regulated shellfish, with consequent higher prices. This has led to illegal harvesting and sale of shellfish on the black market, which can be a health hazard.

Paralytic Shellfish Poisoning

Paralytic shellfish poisoning (PSP) is primarily caused by the consumption of bivalves that have accumulated toxins by feeding on toxic dinoflagellates, single-celled protists found naturally in the sea and inland waters. Saxitoxin is the most virulent of these. In mild cases, PSP causes tingling, numbness, sickness and diarrhoea. In more severe cases, the muscles of the chest wall may be affected leading to paralysis and even death. In 1937, researchers in California established the connection between blooms of these phytoplankton and PSP. The biotoxin remains potent even when the shellfish are well-cooked. In the United States, there is a regulatory limit of 80 µg/g of saxitoxin equivalent in shellfish meat.

Amnesic Shellfish Poisoning

Amnesic shellfish poisoning (ASP) was first reported in eastern Canada in 1987. It is caused by the substance domoic acid found in certain diatoms of the genus *Pseudo-nitzschia*. Bivalves can become toxic when they filter these microalgae out of the water. Domoic acid is a low-molecular weight amino acid that is able to destroy brain cells causing memory loss, gastroenteritis, long-term neurological problems or death. In an outbreak in the western United States in 1993, finfish were also implicated as vectors, and seabirds and mammals suffered neurological symptoms. In the United States and Canada, a regulatory limit of 20 µg/g of domoic acid in shellfish meat is set.

Use in Controlling Pollution

When they live in polluted waters, bivalve molluscs have a tendency to accumulate substances such as heavy metals and persistent organic pollutants in their tissues. This is because they ingest the chemicals as they feed but their enzyme systems are not capable of metabolising them and as a result, the levels build up. This may be a health hazard for the molluscs themselves, and is one for humans who eat them. It also has certain advantages in that bivalves can be used in monitoring the presence and quantity of pollutants in their environment.

There are limitations to the use of bivalves as bioindicators. The level of pollutants found in the tissues varies with species, age, size, time of year and other factors. The quantities of pollutants in the water may vary and the molluscs may reflect past rather than present values. In a study of several bivalve species present in lagoons in Ghana it was found that the findings could be anomalous. Levels of zinc and iron tended to rise in the wet season due to run-off from the galvanized roofing sheets used in many of the houses. Cadmium levels were lower in young animals than in older ones because they were growing so fast that, despite the fact that their bodies were accumulating the metal, the concentration in their tissues reduced. In a study near Vladivostok it was found that the level of pollutants in the bivalve tissues did not always reflect the high levels in the surrounding sediment in such places as harbours. The reason for this was thought to be that the bivalves in these locations did not need to filter so much water as elsewhere because of the water's high nutritional content.

A study of nine different bivalves with widespread distributions in tropical marine waters concluded that the mussel, *Trichomya hirsuta*, most nearly reflected in its tissues the level of heavy metals (Pb, Cd, Cu, Zn, Co, Ni, and Ag) in its environment. In this species there was a linear relationship between the sedimentary levels and the tissue concentration of all the metals except zinc. In the Persian Gulf, the Atlantic pearl-oyster (*Pinctada radiata*) is considered to be a useful bioindicator of heavy metals.

Pacific oyster *Crassostrea gigas* equipped with activity electrodes to follow their daily behaviour

Crushed shells, available as a by-product of the seafood canning industry, can be used to remove pollutants from water. It has been found that, as long as the water is maintained at an alkaline pH, crushed shells will remove cadmium, lead and other heavy metals from contaminated waters by swapping the calcium in their constituent aragonite for the heavy metal, and retaining these pollutants in a solid form. The rock oyster (*Saccostrea cucullata*) has been shown to reduce the levels of copper and cadmium in contaminated waters in the Persian Gulf. The live animals acted as biofilters, selectively removing these metals, and the dead shells also had the ability to reduce their concentration.

Other Uses

Conchology is the scientific study of mollusc shells, but the term conchologist is also sometimes used to describe a collector of shells. Many people pick up shells on the beach or purchase them and display them in their homes. There are many private and public collections of mollusc shells, but the largest one in the world is at the Smithsonian Institution, which houses in excess of 20 million specimens.

Carved shell miniatures

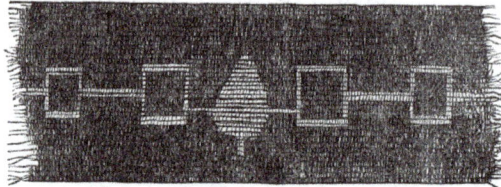
1885 wampum belt

Shells are used decoratively in many ways. They can be pressed into concrete or plaster to make decorative paths, steps or walls and can be used to embellish picture frames, mirrors or other craft items. They can be stacked up and glued together to make ornaments. They can be pierced and threaded onto necklaces or made into other forms of jewellery. Shells have had various uses in the past as body decorations, utensils, scrapers and cutting implements. Carefully cut and shaped shell tools dating back 32,000 years have been found in a cave in Indonesia. In this region, shell technology may have been developed in preference to the use of stone or bone implements, perhaps because of the scarcity of suitable rock materials.

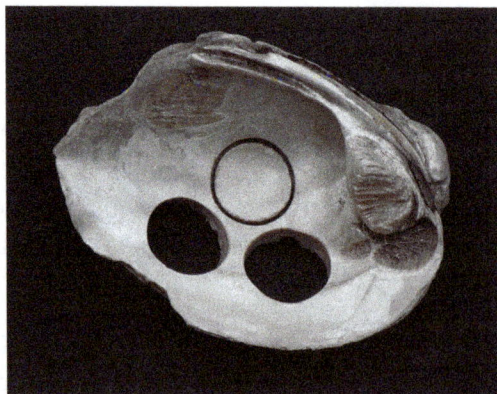
Freshwater mussel shell used for making buttons

The indigenous peoples of the Americas living near the east coast used pieces of shell as wampum. The channeled whelk (*Busycotypus canaliculatus*) and the quahog (*Mercenaria mercenaria*) were used to make white and purple traditional patterns. The shells were cut, rolled, polished and drilled before being strung together and woven into belts. These were used for personal, social and ceremonial purposes and also, at a later date, for currency. The Winnebago Tribe from Wisconsin had numerous uses for freshwater mussels including using them as spoons, cups, ladles and utensils. They notched them to provide knives, graters and saws. They carved them into fish hooks and lures. They incorporated powdered shell into clay to temper their pottery vessels. They used them as scrapers for removing flesh from hides and for separating the scalps of their victims. They used shells as scoops for gouging out fired logs when building canoes and they drilled holes in them and fitted wooden handles for tilling the ground.

Buttons have traditionally been made from a variety of freshwater and marine shells. At first they were used decoratively rather than as fasteners and the earliest known example dates back five thousand years and was found at Mohenjo-daro in the Indus Valley.

Carved nacre in a 16th-century altarpiece

Sea silk is a fine fabric woven from the byssus threads of bivalves, particularly the pen shell (*Pinna nobilis*). It used to be produced in the Mediterranean region where these shells are endemic. It was an expensive fabric and overfishing has much reduced populations of the pen shell. There is mention in the Greek text on the Rosetta Stone (196 BCE) of this cloth being used to pay taxes.

Crushed shells are added as a calcareous supplement to the diet of laying poultry. Oyster shell and cockle shell are often used for this purpose and are obtained as a by-product from other industries.

Pearls and Mother-of-pearl

Mother-of-pearl or nacre is the naturally occurring lustrous layer that lines some mollusc shells. It is used to make pearl buttons and in artisan craftwork to make organic jewellery. It has traditionally been inlaid into furniture and boxes, particularly in China. It has been used to decorate musical instruments, watches, pistols, fans and other products. The import and export of goods made with nacre are controlled in many countries under the International Convention of Trade in Endangered Species of Wild Fauna and Flora.

A pearl is created in the mantle of a mollusk when an irritant particle is surrounded by layers of nacre. Although most bivalves can create pearls, oysters in the family Pteriidae and freshwater mussels in the families Unionidae and Margaritiferidae are the main source of commercially available pearls because the calcareous concretions produced by most other species have no lustre. Finding pearls inside oysters is a very chancy business as hundreds of shells may need to be pried open before a single pearl can be found. Most pearls are now obtained from cultured shells where an irritant substance has been purposefully introduced to induce the formation of a pearl. A "mabe" (irregular) pearl can be grown by the insertion of an implant, usually made of plastic, under a flap of the mantle and next to the mother-of-pearl interior of the shell. A more difficult procedure is the grafting of a piece of oyster mantle into the gonad of an adult specimen together with the insertion of a shell bead nucleus. This produces a superior, spherical pearl. The animal can be opened to extract the pearl after about two years and reseeded so that it produces another pearl. Pearl oyster farming and pearl culture is an important industry in Japan and many other countries bordering the Indian and Pacific Oceans.

Symbolism

The scallop is the symbol of St James and is called *Coquille Saint-Jacques* in French. It is an emblem carried by pilgrims on their way to the shrine of Santiago de Compostela in Galicia. The shell became associated with the pilgrimage and came to be used as a symbol showing hostelries along the route and later as a sign of hospitality, food and lodging elsewhere.

Roman myth has it that Venus, the goddess of love, was born in the sea and emerged accompanied by fish and dolphins, with Botticelli depicting her as arriving in a scallop shell. The Romans revered her and erected shrines in her honour in their gardens, praying to her to provide water and verdant growth. From this, the scallop and other bivalve shells came to be used as a symbol for fertility. Its depiction is used in architecture, furniture and fabric design and it is the logo of Royal Dutch Shell, the global oil and gas company.

Bivalvian Taxonomies

For the past two centuries no consensus has existed on bivalve phylogeny from the many classifications developed. In earlier taxonomic systems, experts used a single characteristic feature for their classifications, choosing among shell morphology, hinge type or gill type. Conflicting naming schemes proliferated due to these taxonomies based on single organ systems. One of the most widely accepted systems was that put forward by Norman D. Newell in Part N of the *Treatise on Invertebrate Paleontology*, which employed a classification system based on general shell shape, microstructures and hinge configuration. Because features such as hinge morphology, dentition, mineralogy, shell morphology and shell composition change slowly over time, these characteristics can be used to define major taxonomic groups.

Mussels in the intertidal zone in Cornwall, England

Since the year 2000, taxonomic studies using cladistical analyses of multiple organ systems, shell morphology (including fossil species) and modern molecular phylogenetics have resulted in the drawing up of what experts believe is a more accurate phylogeny of the Bivalvia. Based upon these studies, a new proposed classification system for the Bivalvia was published in 2010 by Bieler, Carter & Coan. In 2012, this new system was adopted by the World Register of Marine Species (WoRMS) for the classification of the Bivalvia. Some experts still maintain that Anomalodesmacea should be considered a separate subclass, whereas the new system treats it as the order Anomalodesmata, within the subclass Heterodonta. Molecular phylogenetic work continues, further clarifying which Bivalvia are most closely related and thus refining the classification.

Fossil gastropod and attached mytilid bivalves in a Jurassic limestone
(Matmor Formation) in southern Israel

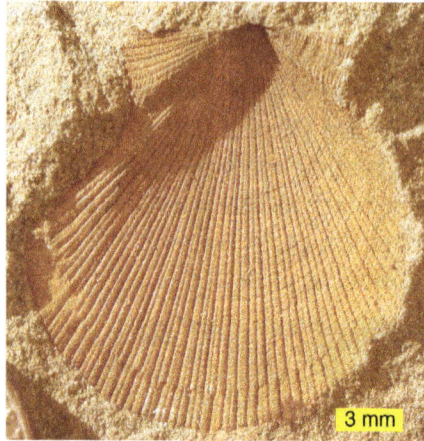

Aviculopecten subcardiformis; a fossil of an extinct scallop from the Logan Formation of Wooster, Ohio (external mold)

Practical Taxonomy of R.C. Moore

R.C. Moore, in Moore, Lalicker, and Fischer, 1952, *Invertebrate Fossils*, gives a practical and useful classification of pelecypods (Bivalvia) even if somewhat antiquated, based on shell structure, gill type, and hinge teeth configuration. Subclasses and orders given are:

Subclass:Prionodesmacea

Order

- Paleoconcha

- Taxodonta: Many teeth (e.g. order Nuculida)

- Schizodonta: Big bifurcating teeth (e.g. *Trigonia* spp.)

- Isodonta: Equal teeth (e.g. *Spondylus* spp.)

- Dysodonta: Absent teeth and ligaments joins the valves.

Subclass:Teleodesmacea

Order

- Heterodonta: Different teeth (e.g. family Cardiidae). [Lower Ordovician – Recent]

- Pachydonta: Large, different, deformed teeth (e.g. rudist spp.). [Late Jurassic – Upper Cretaceous]

- Desmodonta: Hinge-teeth absent or irregular with ligaments (e.g. family Anatinidae).

- Prionodesmacea have a prismatic and nacreous shell structure, separated mantle lobes, poorly developed siphons, and hinge teeth that are lacking or unspecialized. Gills range from protobranch to eulamellibranch. Teleodesmacea on the other hand have a porcelanous and partly nacreous shell structure; Mantle lobes that are generally connected, well developed siphons, and specialized hinge teeth. In most, gills are eulamellibranch.

1935 Taxonomy

In his 1935 work *Handbuch der systematischen Weichtierkunde* (Handbook of Systematic Malacology), Johannes Thiele introduced a mollusc taxonomy based upon the 1909 work by Cossmann and Peyrot. Thiele's system divided the bivalves into three orders. *Taxodonta* consisted of forms that had taxodont dentition, with a series of small parallel teeth perpendicular to the hinge line. *Anisomyaria* consisted of forms that had either a single adductor muscle or one adductor muscle much smaller than the other. *Eulamellibranchiata* consisted of forms with ctenidial gills. The Eulamellibranchiata was further divided into four suborders: *Schizodonta, Heterodonta, Adapedonta* and *Anomalodesmata*.

Mussel

Mussel is the common name used for members of several families of bivalve molluscs, from saltwater and freshwater habitats. These groups have in common a shell whose outline is elongated and asymmetrical compared with other edible clams, which are often more or less rounded or oval.

The word "mussel" is most frequently used to mean the edible bivalves of the marine family Mytilidae, most of which live on exposed shores in the intertidal zone, attached by means of their strong byssal threads ("beard") to a firm substrate. A few species (in the genus *Bathymodiolus*) have colonised hydrothermal vents associated with deep ocean ridges.

In most marine mussels the shell is longer than it is wide, being wedge-shaped or asymmetrical. The external colour of the shell is often dark blue, blackish, or brown, while the interior is silvery and somewhat nacreous.

The common name "mussel" is also used for many freshwater bivalves, including the freshwater pearl mussels. Freshwater mussel species inhabit lakes, ponds, rivers, creeks, canals, and they are classified in a different subclass of bivalves, despite some very superficial similarities in appearance.

Freshwater zebra mussels and their relatives in the family Dreissenidae are not related to previously mentioned groups, even though they resemble many *Mytilus* species in shape, and live attached to rocks and other hard surfaces in a similar manner, using a byssus. They are classified with the Heterodonta, the taxonomic group which includes most of the bivalves commonly referred to as "clams".

General Anatomy

Marine blue mussel, *Mytilus edulis*, showing some of the inner anatomy. The white posterior adductor muscle is visible in the upper image, and has been cut in the lower image to allow the valves to open fully.

Flight around a 3D-Rendering of a µCT-Scan of a young Mytilus that is almost completely covered with Balanidae (barnacles). Resolution of the scan is 29µm/Voxel.

The mussel's external shell is composed of two hinged halves or "valves". The valves are joined together on the outside by a ligament, and are closed when necessary by strong internal muscles (anterior and posterior adductor muscles). Mussel shells carry out a variety of functions, including support for soft tissues, protection from predators and protection against desiccation.

The shell has three layers. In the pearly mussels there is an inner iridescent layer of nacre (mother-of-pearl) composed of calcium carbonate, which is continuously secreted by the mantle; the prismatic layer, a middle layer of chalky white crystals of calcium carbonate in a protein matrix; and the periostracum, an outer pigmented layer resembling a skin. The periostracum is composed of a protein called conchin, and its function is to protect the prismatic layer from abrasion and dissolution by acids (especially important in freshwater forms where the decay of leaf materials produces acids).

Like most bivalves, mussels have a large organ called a foot. In freshwater mussels, the foot is large, muscular, and generally hatchet-shaped. It is used to pull the animal through the substrate (typically sand, gravel, or silt) in which it lies partially buried. It does this by repeatedly advancing the foot through the substrate, expanding the end so it serves as an anchor, and then pulling the rest of the animal with its shell forward. It also serves as a fleshy anchor when the animal is stationary.

In marine mussels, the foot is smaller, tongue-like in shape, with a groove on the ventral surface which is continuous with the byssus pit. In this pit, a viscous secretion is exuded, entering the groove and hardening gradually upon contact with sea water. This forms extremely tough, strong, elastic, byssal threads that secure the mussel to its substrate allowing it to remain sessile in areas of high flow. The byssal thread is also sometimes used by mussels as a defensive measure, to tether predatory molluscs, such as dog whelks, that invade mussel beds, immobilising them and thus starving them to death.

In cooking, the byssus of the mussel is known as the "beard" and is removed before the mussels are prepared.

Life Habits

A *Mytilus* with its byssus clearly showing, at Ocean Beach, San Francisco, California

A starfish consuming a mussel in Northern California

Feeding

Both marine and freshwater mussels are filter feeders; they feed on plankton and other microscopic sea creatures which are free-floating in seawater. A mussel draws water in through its incurrent siphon. The water is then brought into the branchial chamber by the actions of the cilia located on the gills for ciliary-mucus feeding. The wastewater exits through the excurrent siphon. The labial palps finally funnel the food into the mouth, where digestion begins.

Marine mussels are usually found clumping together on wave-washed rocks, each attached to the rock by its byssus. The clumping habit helps hold the mussels firm against the force of the waves. At low tide mussels in the middle of a clump will undergo less water loss because of water capture by the other mussels.

Reproduction

Both marine and freshwater mussels are gonochoristic, with separate male and female individuals. In marine mussels, fertilization occurs outside the body, with a larval stage that drifts for three weeks to six months, before settling on a hard surface as a young mussel. There, it is capable of moving slowly by means of attaching and detaching byssal threads to attain a better life position.

Freshwater mussels reproduce sexually. Sperm is released by the male directly into the water and enters the female via the incurrent siphon. After fertilization, the eggs develop into a larval stage called a glochidium (plural glochidia), which temporarily parasitizes fish, attaching themselves to the fish's fins or gills. Prior to their release, the glochidia grow in the gills of the female mussel where they are constantly flushed with oxygen-rich water. In some species, release occurs when a fish attempts to attack the mussel's minnow or other mantle flaps shaped like prey; an example of aggressive mimicry.

Glochidia are generally species-specific, and will only live if they find the correct fish host. Once the larval mussels attach to the fish, the fish body reacts to cover them with cells forming a cyst, where the glochidia remain for two to five weeks (depending on temperature). They grow, break free from the host, and drop to the bottom of the water to begin an independent life.

Predators

Marine mussels are eaten by humans, starfish, seabirds, and by numerous species of predatory marine gastropods in the family Muricidae, such as the dog whelk, *Nucella lapillus.* Freshwater mussels are eaten by otters, raccoons, ducks, baboons, humans (off the coast of South Africa) and geese, although the main cause of mortality is starfish. To resist predation from starfish, the mussel can physically make its shell thicker hence can become stronger.

Distribution and Habitat

Marine mussels are abundant in the low and mid intertidal zone in temperate seas globally. Other species of marine mussel live in tropical intertidal areas, but not in the same huge numbers as in temperate zones.

Mussels completely covering rocks in intertidal zone, in Dalian, Liaoning Province, China

Certain species of marine mussels prefer salt marshes or quiet bays, while others thrive in pounding surf, completely covering wave-washed rocks. Some species have colonized abyssal depths near hydrothermal vents. The South African white mussel exceptionally doesn't bind itself to rocks but burrows into sandy beaches extending two tubes above the sand surface for ingestion of food and water and exhausting wastes.

Freshwater mussels inhabit permanent lakes, rivers, canals and streams throughout the world except in the polar regions. They require a constant source of cool, clean water. They prefer water with a substantial mineral content, using calcium carbonate to build their shells.

Aquaculture

Mussel dredgers

In 2005, China accounted for 40% of the global mussel catch according to a FAO study. Within Europe, where mussels have been cultivated for centuries, Spain remained the industry leader. Aquaculture of mussels in North America began in the 1970s. In the US, the northeast and northwest have significant mussel aquaculture operations, where *Mytilus edulis* (blue mussel) is most commonly grown. While the mussel industry in the US has increased, in North America, 80% of cultured mussels are produced in Prince Edward Island in Canada. In Washington State, an estimated 2.9M pounds of mussels were harvested in 2010, valued at roughly $4.3M.

Bouchots are marine pilings for growing mussels, here shown at an agricultural fair.

Bamboo is used for mussel breeding and propagation (Abucay, Bataan, Philippines).

Culture Methods

Freshwater mussels are used as host animals for the cultivation of freshwater pearls. Some species of marine mussel, including the Blue mussel (*Mytilus edulis*) and the New Zealand green-lipped mussel (*Perna canaliculus*), are also cultivated as a source of food.

In some areas of the world, mussel farmers collect naturally occurring marine mussel seed for transfer to more appropriate growing areas, however, most North American mussel farmers rely on hatchery-produced seed. Growers typically purchase seed after it has set (about 1mm in size) or after it has been nursed in upwellers for 3-6 additional weeks and is 2-3mm. The seed is then typically reared in a nursery environment, where it is transferred to a material with a suitable surface for later relocation to the growing area. After about three months in the nursery, mussel seed is "socked" (placed in a tube-like mesh material) and hung on longlines or rafts for grow-out. Within a few days, the mussels migrate to the outside of the sock for better access food sources in the water column. Mussels grow quickly and are usually ready for harvest in less than two years. Unlike other cultured bivalves, mussels use byssus threads (beard) to attach themselves to any firm substrate, which makes them suitable for a number of culture methods.

There are a variety of techniques for growing mussels.

- Bouchot culture: Intertidal growth technique, or bouchot technique: pilings, known in French as bouchots, are planted at sea; ropes, on which the mussels grow, are tied in a spiral on the pilings; some mesh netting prevents the mussels from falling away. This method needs an extended tidal zone.

- On-bottom culture: On-bottom culture is based on the principle of transferring mussel seed (spat) from areas where they have settled naturally to areas where they can be placed in lower densities to increase growth rates, facilitate harvest, and control predation (Mussel farmers must remove predators and macroalgae during the growth cycle).

- Raft culture: Raft culture is a commonly used method throughout the world. Lines of rope mesh socks are seeded with young mussels and suspended vertically from a raft. The specific length of the socks depends on depth and food availability.

- Longline culture (rope culture): Mussels are cultivated extensively in New Zealand, where the most common method is to attach mussels to ropes which are hung from a rope back-bone supported by large plastic floats. The most common species cultivated in New Zealand is the New Zealand green-lipped mussel. Longline culture is the most recent development for mussel culture and are often used as an alternative to raft culture in areas that are more exposed to high wave energy. A long-line is suspended by a series of small anchored floats and ropes or socks of mussels are then suspended vertically from the line.

Harvest

In roughly 12–15 months, mussels reach marketable size (40mm) and are ready for harvest. Harvesting methods depend on the grow-out area and the rearing method being used. Dredges are currently used for on-bottom culture. Mussels grown on wooden poles can be harvested by hand or with a hydraulic powered system. For raft and longline culture, a platform is typically lowered under the mussel lines, which are then cut from the system and brought to the surface and dumped into containers on a nearby vessel. After harvest, mussels are typically placed in seawater tanks to rid them of impurities before marketing.

Medical

Byssal threads, used to anchor mussels to substrates, are now recognized as superior bonding agents. A number of studies have investigated mussel "glues" for industrial and surgical applications.

Additionally byssal threads have provided insight into the construction of artificial tendons.

Environmental Applications

Mussels are widely used as bio-indicators to monitor the health of aquatic environments in both fresh water and the marine environments. They are particularly useful since they are distributed worldwide and they are sessile. These characteristics ensure that they are representative of the environment where they are sampled or placed. Their population status or structure, physiology, behaviour or the level of contamination with elements or compounds can indicate the status of the ecosystem.

Mussels and Nutrient Mitigation

Marine nutrient bioextraction is the practice of farming and harvesting marine organisms such as shellfish and seaweed for the purpose of reducing nutrient pollution. Mussels and other bivalve shellfish consume phytoplankton containing nutrients such as nitrogen (N) and phosphorus (P). On average, one live mussel is 1.0% N and 0.1% P. When the mussels are harvested and removed, these nutrients are also removed from the system and recycled in the form of seafood or mussel biomass, which can be used as an organic fertilizer or animal feed-additive. These ecosystem services provided by mussels are of particular interest to those hoping to mitigate excess anthropogenic marine nutrients, particularly in eutrophic marine systems. While mussel aquaculture is actually promoted in some countries such as Sweden as a water management strategy to address coastal eutrophication, mussel farming as a nutrient mitigation tool is still in its infancy in most parts of the world. Ongoing efforts in the Baltic Sea (Denmark, Sweden, Germany, Poland) and Long Island Sound and Puget Sound in the U.S. are currently examining nutrient uptake, cost-effectiveness, and potential environmental impacts of mussel farming as a means to mitigate excess nutrients and complement traditional wastewater treatment programs.

Conservation

Freshwater Mussels

In the United States and Canada, areas home to the most diverse freshwater mussel fauna in the world, there are 297 known freshwater mussel taxa. Of the 297 known species, 213 (71.7%) taxa are listed as endangered, threatened, of special concern. The main factors contributing to the decline of freshwater mussels include destruction from dams, increased siltation, channel modification, and the introduction of invasive species like the Zebra mussel.

As Food

Humans have used mussels as food for thousands of years. About 17 species are edible, of which the most commonly eaten are *Mytilus edulis, M. galloprovincialis, M. trossellus* and *Perna canaliculus.*

Freshwater mussels nowadays are generally considered to be unpalatable, though the native peoples in North America ate them extensively. During the second World War in the United States, mussels were commonly served in diners. This was due to the unavailability of red meat related to wartime rationing.

In Belgium, the Netherlands, and France, mussels are consumed with french fries ("mosselen met friet" or "moules-frites") or bread. In Belgium, mussels are sometimes served with fresh herbs and flavorful vegetables in a stock of butter and white wine. Fries and Belgian beer sometimes are accompaniments. In the Netherlands, mussels are sometimes served fried in batter or breadcrumbs, particularly at take-out food outlets or informal settings. In France, the Éclade des Moules, or, locally, Terré de Moules, is a mussel bake that can be found along the beaches of the Bay of Biscay.

In Italy, mussels are mixed with other sea food, they are consumed often steam cooked (most popular), sometimes with white wine, herbs, and served with the remaining water and some lemon. In Spain, they are consumed mostly steam cooked, sometimes boiling white wine, onion and herbs, and served with the remaining water and some lemon. They can also be eaten as "tigres", a sort of croquette using the mussel meat, shrimps and other pieces of fish in a thick bechamel then breaded and fried in the clean mussel shell. They are used in other sort of dishes such as rices or soups or commonly eaten canned in a pickling brine made of oil, vinegar, peppercorns, bay leaves and paprika.

In Turkey, mussels are either covered with flour and fried on shishs ('midye tava'), or filled with rice and served cold ('midye dolma') and are usually consumed after alcohol (mostly raki or beer).

They are used in Ireland boiled and seasoned with vinegar, with the "bray" or boiling water as a supplementary hot drink.

In Cantonese cuisine, mussels are cooked in a broth of garlic and fermented black bean. In New Zealand, they are served in a chili or garlic-based vinaigrette, processed into fritters and fried, or used as the base for a chowder.

In India, mussels are popular in Kerala, Maharashtra, Karnataka-Bhatkal, and Goa. They are either prepared with drumsticks, breadfruit or other vegetables, or filled with rice and coconut paste with spices and served hot. Fried mussels of north Kerala especially in kozhikode are a spicy, favored delicacy. In coastal Karnataka Beary's prepare special rice ball stuffed with spicy fried mussels and steamed locally known as "pachilede pindi".

Preparation

Mussels can be smoked, boiled, steamed, roasted, barbecued or fried in butter or vegetable oil. As with all shellfish, except shrimp, mussels should be checked to ensure they are still alive just before they are cooked; enzymes quickly break down the meat and make them unpalatable or poisonous after dying or uncooked. Some mussels might contain toxins. A simple criterion is that live mussels, when in the air, will shut tightly when disturbed. Open, unresponsive mussels are dead, and must be discarded. Unusually heavy, wild-caught, closed mussels may be discarded as they may contain only mud or sand. (They can be tested by slightly opening the shell halves.) A thorough rinse in water and removal of "the beard" is suggested. Mussel shells usually open when cooked, revealing the cooked soft parts. Historically, it has been believed that after cooking all the mussels should have opened and those that have not are not safe to eat and should be discarded. However, according to marine biologist Nick Ruello, this advice may have arisen from an old, poorly researched cookbook's advice, which has now become an assumed truism for all shellfish. Ruello found 11.5% of all mussels failed to open during cooking, but when forced open, 100% were "both adequately cooked and safe to eat."

Moules frites

A mussel dish with cherry tomatoes and croutons

Although mussels are valued as food, mussel poisoning due to toxic planktonic organisms can be a danger along some coastlines. For instance, mussels should be avoided along the west coast of the United States during the warmer months. This poisoning is usually due to a bloom of dinoflagellates (red tides), which contain toxins. The dinoflagellates and their toxin are harmless to mussels, even when concentrated by the mussel's filter feeding, but if the mussels are consumed by humans, the concentrated toxins cause serious illness, such as paralytic shellfish poisoning. A person affected in this way after eating mussels is said to be *mussel(l)ed*.

- Soft clam: *Mya arenaria*

- Atlantic surf clam: *Spisula solidissima*

- Ocean quahog: *Arctica islandica*

- Pacific razor clam: *Siliqua patula*

- Pismo clam: *Tivela stultorum* (8 inch shell on display at the Pismo Beach Chamber of Commerce)

- Geoduck: *Panopea abrupta* or *Panope generosa* (largest burrowing clam in the world)

- Atlantic jackknife clam: *Ensis directus*

- Lyrate Asiatic hard clam: Meretrix lyrata

- Ark clams, family Arcidae (most popular in Indonesia and Singapore)

Not usually considered edible:

- Nut clams or pointed nut clams, family Nuculidae

- Duck clams or trough shells, family Mactridae

- Marsh clams, family Corbiculidae

- File clams, family Limidae

- Giant clam: *Tridacna gigas*

- Asian or Asiatic clam: genus *Corbicula*

- Peppery furrow shell: *Scrobicularia plana*

Freshwater Pearl Mussel

The freshwater pearl mussel, scientific name *Margaritifera margaritifera*, is an endangered species of freshwater mussel, an aquatic bivalve mollusc in the family Margaritiferidae.

Although the name "freshwater pearl mussel" is often used for this species, other freshwater mussel species can also create pearls and some can also be used as a source of mother of pearl. In fact, most cultured pearls today come from *Hyriopsis* species in Asia, or *Amblema* species in North America, both members of the related family Unionidae; pearls are also found within species in the genus *Unio*.

The interior of the shell of *Margaritifera margaritifera* has thick nacre (the inner mother of pearl layer of the shell). This species is capable of making fine-quality pearls, and was historically exploited in the search for pearls from wild sources. In recent times, the Russian malacologist Valeriy Zyuganov received worldwide reputation after he discovered that the pearl mussel exhibited negligible senescence and he determined that it had a maximum lifespan of 210–250 years. The data of V.V. Zyuganov have been confirmed by the Finnish malacologists and gained general acceptance.

Subspecies

Subspecies within the species *Margaritifera magaritifera* include:

- *Margaritifera margaritifera margaritifera* (Linnaeus, 1758)

- *Margaritifera margaritifera parvula* (Haas, 1908)

- *Margaritifera margaritifera durrovensis* Phillips, 1928 - critically endangered subspecies in Ireland. Synonym: *Margaritifera durrovensis*. This subspecies is mentioned in annexes II and V of Habitats Directive as *Margaritifera durrovensis*.

Description

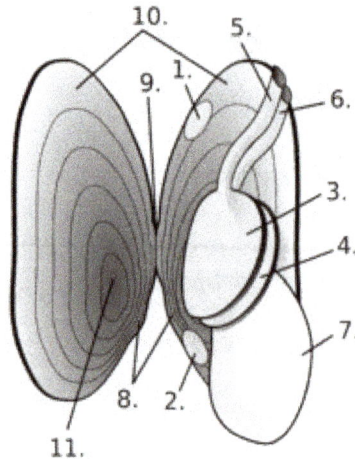

The anatomy of *Margaritifera margaritifera*

1. Posterior adductor muscle

2. Anterior adductor muscle

3. Frontal gill

4. Back gill

5. Exhalant aperture

6. Inhalant aperture

7. Foot

8. Pseudotooth

9. The hingeline and ligament

10. Mantle

11. The shell's thickest part, the umbo

The freshwater pearl mussel is one of the longest-living invertebrates in existence. The oldest known specimen in Europe was caught in 1993 in Estonia when it was 134 years old.

Like all bivalve molluscs, the freshwater pearl mussel has a shell consisting of two parts that are hinged together, which can be closed to protect the animal's soft body within. The shell is large, heavy and elongated, typically yellowish-brown in colour when young and becoming darker with age. Older parts of the shell often appear corroded, an identifying feature of this mussel species. The inner surface of the shell is pearl white, sometimes tinged with attractive iridescent colours. Like all molluscs, the freshwater pearl mussel has a muscular 'foot'; this very large, white foot enables the mussel to move slowly and bury itself within the bottom substrate of its freshwater habitat.

Distribution

Group of live *Margaritifera margaritifera* in a river bed in Sweden

The native distribution of this species is Holarctic. The freshwater pearl mussel can be found on both sides of the Atlantic, from the Arctic and temperate regions of western Russia, through Europe to northeastern North America.

- North America: eastern Canada and New England in the United States' Northeast

- Europe, including:

    o   Austria - estimated total population of 70 000 individuals in Mühlviertel (declining) and in Waldviertel (some recruitment), in the states of Upper and Lower Austria, respectively.

o   Belgium

o   Czech Republic - critically endangered (CR). In Bohemia, probably locally extinct in Moravia. Its Conservation status in 2004-2006 is bad (U2) in report for European commission in accordance with Habitats Directive.

o   Denmark

o   Estonia

o   Fennoscandia - vulnerable in Finland and Norway, endangered in Sweden. Very rare in southern Finland, more common in the north. Widespread but not common in Norway; Norway is considered to host a large proportion of the European stock. Rare in Sweden. Also in Kola Peninsula and Karelia (Russia).

o   France

o   Germany - critically endangered (*vom Aussterben bedroht*). Listed as strictly protected species in annex 1 in Bundesartenschutzverordnung.

o   Great Britain. More than half the world's recruiting population exists in Scotland with populations in more than 50 rivers, mainly in the Highlands, although illegal harvesting has seriously affected their survival. 75% of sites surveyed in 2010 had suffered "significant and lasting criminal damage" and in response the police and Scottish Natural Heritage have launched a campaign to protect the species. This species has been fully protected in the United Kingdom under the Wildlife and Countryside Act 1981 since 1998 and partly protected according to section 9(1) since 1991.

o   Iberian Peninsula (Portugal and Spain)

o   Ireland. The Cladagh (Swanlinbar) river contains one of the largest populations surviving in northern Ireland, estimated minimum 10,000, confined to a 6 km stretch of undisturbed river in the middle section.

o   Luxembourg

o   Latvia

o   Lithuania - extinct

o   Poland - extinct

o   Russian Federation - in the rivers of the White Sea basin of the Arkhangelsk and Murmansk Regions. It is east border of the area of distribution *M. margaritifera*.

Habitat

Clean, fast-flowing streams and rivers are required for the freshwater pearl mussel, where it lives buried or partly buried in fine gravel and coarse sand, generally in water at depths between 0.5 and 2 metres, but sometimes at greater depths. Clean gravel and sand is essential, particularly for juvenile freshwater pearl mussels, for if the stream or river bottom becomes clogged with silt, they cannot obtain oxygen and will die. Also essential is the presence of a healthy population of salmonids, a group of fish including salmon and trout, on which the freshwater pearl mussel relies for part of its life cycle.

Lifecycle

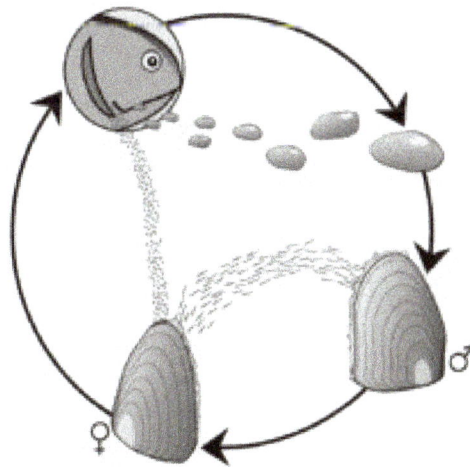

life cycle

Capable of living for up to 130 years, the freshwater pearl mussel begins life as a tiny larva, measuring just 0.6 to 0.7 millimetres long, which is ejected into the water from an adult mussel in a mass of one to four million other larvae. This remarkable event takes place over just one to two days, sometime between July and September. The larvae, known as glochidia, resemble tiny mussels, but their minute shells are held open until they snap shut on a suitable host. The host of freshwater pearl mussel larvae are juvenile fish from the salmonid family, which includes the Atlantic salmon and sea trout. The chances of a larva encountering a suitable fish is very low, and thus nearly all are swept away and die; only a few are inhaled by an Atlantic salmon or sea trout, where they snap shut onto the fish's gills.

Attached to the gills of a fish, the glochidia live and grow in this oxygen-rich environment until the following May or June, when they drop off. The juvenile must land on clean gravely or sandy substrates if it is to successfully grow. Attached to the substrate, juvenile freshwater pearl mussels typically burrow themselves completely into the sand or gravel, while adults are generally found with a third of their shell exposed. Should they become dislodged, freshwater pearl mussels can rebury themselves, and are also capable of moving slowly across sandy sediments, using their large, muscular foot.

The freshwater pearl mussel grows extremely slowly, inhaling water through exposed siphons, and filtering out tiny organic particles on which it feeds. It is thought that in areas where this species was once abundant, this filter feeding acted to clarify the water, benefiting other species which inhabited the rivers and streams. Maturity is reached at an age of 10 to 15 years, followed by a reproductive period of over 75 years in which about 200 million larvae can be produced. In early summer each year, around June and July, male freshwater pearl mussels release sperm into the water, where they are inhaled by female mussels. Inside the female, the fertilized eggs develop in a pouch on the gills for several weeks, until temperature or other environmental cues trigger the female to release the larvae into the surrounding water.

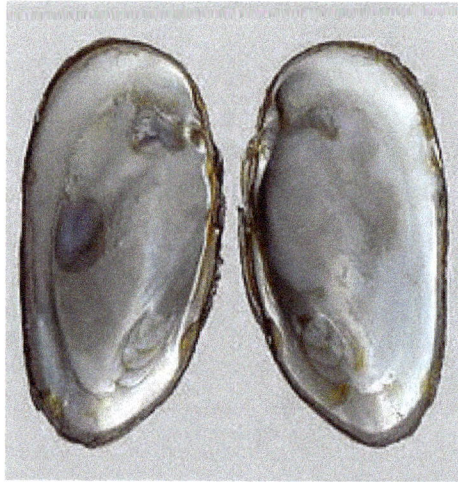

The interior of the shell of *Margaritifera margaritifera*, showing the nacre

Threats and Conservation

Once the most abundant bivalve mollusc in ancient rivers around the world, numbers of the freshwater pearl mussel are now declining in all countries and this species is nearly extinct in many areas. The causes of this decline are not fully understood, but alteration and degradation of its freshwater habitat undoubtedly plays a central role. The negative impacts humans have on rivers and streams come from a wide range of activities such as river regulation, drainage, sewage disposal, dredging, and water pollution, including the introduction of excess nutrients. Anything that affects the abundance of the fish hosts will also affect the freshwater pearl mussel; for example, the introduction of exotic fish species, such as the rainbow trout, reduce the number of native fish hosts. Introduced species are also directly affecting the freshwater pearl mussel; the invasion of the zebra mussel (*Dreissena polymorpha*), which has been spread to new locations by being transported on the bottom of boats or in ballast waters, has impacted freshwater pearl mussel populations in all countries it has invaded.

The freshwater pearl mussel, which is completely protected in all European countries, has been the focus of a significant amount of conservation efforts. Measures have included the transfer of adult mussels to areas where it had gone extinct, the culture of juvenile mussels, and the release of juvenile trout, which have been infected with glochidia, into small rivers, but mainly the freshwater pearl mussel has benefited from habitat restoration projects in some areas. Due to the essential role salmonid fish play in the life of the freshwater pearl mussel, the conservation of salmon and trout is also central in the survival of this endangered freshwater mussel.

Freshwater Bivalve

Freshwater bivalves are one kind of freshwater molluscs, along with freshwater snails. They are bivalves which live in freshwater, as opposed to saltwater, the main habitat type for bivalves.

The majority of species of bivalve molluscs live in the sea, but in addition, a number of different families live in freshwater (and in some cases also in brackish water). These families belong to two different evolutionary lineages (freshwater mussels and freshwater clams), and the two groups are not closely related.

Freshwater bivalves live in many types of habitat, ranging from small ditches and ponds, to lakes, canals, rivers, and swamps.

Species in the two groups vary greatly in size. Some of the pea clams (*Pisidium* species) have an adult size of only 3 mm. In contrast, one of the largest species of freshwater bivalves is the swan mussel, in the family Unionidae; it can grow to a length of 20 cm, and usually lives in lakes or slow rivers.

Freshwater pearl mussels are economically important as a source of freshwater pearls and mother of pearl.

Freshwater Snail

Bithynia tentaculata, a small freshwater gastropod in the family Bithyniidae

Pomacea insularum, an apple snail

Freshwater snails are gastropod mollusks which live in freshwater. There are many different families. They are found throughout the world in various habitats, ranging from ephemeral pools to the largest lakes, and from small seeps and springs to major rivers. The great majority of freshwater gastropods have a shell, with very few exceptions. Some groups of snails that live in freshwater respire using gills, whereas other groups need to reach the surface to breathe air. Most feed on algae, but many are detritivors and some are filter feeders.

According to a 2008 review of the taxonomy, there are about 4,000 species of freshwater gastropods (3,795-3,972).

Planorbella trivolvis, an air-breathing ramshorn snail

At least 33–38 independent lineages of gastropods have successfully colonized freshwater environments. It is not possible to quantify the exact number of these lineages yet, because they have yet to be clarified within the Cerithioidea. From six to eight of these independent lineages occur in North America.

Taxonomy

2005 Taxonomy

The following cladogram is an overview of the main clades of gastropods based on the taxonomy of Bouchet & Rocroi (2005), with families that contain freshwater species marked in boldface: (Some of the highlighted families consist entirely of freshwater species, but some of them also contain, or even mainly consist of, marine species.)

2010 Taxonomy

The following cladogram is an overview of the main clades of gastropods based on the taxonomy of Bouchet & Rocroi (2005), modified after Jörger et al. (2010) and simplified with families that contain freshwater species marked in boldface: (Marine gastropods (Siphonarioidea, Sacoglossa, Amphiboloidea, Pyramidelloidea) are not depicted within Panpulmonata for simplification. Some of these highlighted families consist entirely of freshwater species, but some of them also contain, or even mainly consist of, marine species.)

Neritimorpha

The Neritimorpha are a group of primitive "prosobranch" gilled snails which have a shelly operculum.

- Neritiliidae, 5 extant freshwater species

- Neritidae, largely confined to the tropics, also the rivers of Europe, family includes the marine "nerites". There are about 110 extant freshwater species.

Family Neritidae, shells of *Theodoxus fluviatilis*.

Family Neritidae, *Neritina natalensis*

Caenogastropoda

The Caenogastropoda are a large group of gilled operculate snails, which are largely marine. In freshwater habitats there are ten major families of caenogastropods, as well as several other families of lesser importance:

Architaenioglossa

- Ampullariidae, an exclusively freshwater family that is largely tropical and includes the large "apple snails" kept in aquaria. 105-170 species.

- Viviparidae, medium to large snails, live-bearing, commonly referred to as "mystery snails". Worldwide except South America, and everywhere confined to fresh waters. 125-150 species.

Family Ampullariidae, *Pomacea bridgesii*.

Family Viviparidae, *Viviparus viviparus*.

Sorbeoconcha

- Melanopsidae, family native to rivers draining to the Mediterranean, also Middle East, and some South Pacific islands. About 25-50 species.

- Pachychilidae - 165-225 species. native to South and Central America. Formerly included with the Pleuroceridae by many authors.

- Paludomidae - about 100 species in south Asia, diverse in African Lakes, and Sri Lanka. Formerly classified with the Pleuroceridae by some authors.

- Pleuroceridae, abundant and diverse in eastern North America, largely high-spired snails of small to large size. About 150 species.

- Semisulcospiridae, - primarily eastern Asia, Japan, also the *Juga* snails of northwestern North America. Formerly included with the Pleuroceridae. About 50 species.

- Thiaridae, high-spired parthenogenic snails of the tropics, includes those referred to as "trumpet snails" in aquaria. About 110 species.

Family Pleuroceridae, *Io fluvialis*.

Family Semisulcospiridae, *Semisulcospira kurodai*.

Littorinimorpha

- Littorinidae - 9 species in the genus *Cremnoconchus* are freshwater living in streams and waterfalls. Other species are marine.

- Amnicolidae - about 200 species.

- Assimineidae - about 20 freshwater species, other are marine

- Bithyniidae, small snails, native to Eastern Hemisphere. About 130 species.

- Cochliopidae - 246 species.

- Helicostoidae, the only species *Helicostoa sinensis* lives in China.

- Hydrobiidae, small to very small snails found worldwide. About 1250 freshwater species other are marine.

Clea helena, family Buccinidae.

- Lithoglyphidae - about 100 species.

- Moitessieriidae - 55 species.

- Pomatiopsidae, small amphibious snails scattered worldwide, most diverse in eastern and Southeast Asia. About 170 species.

- Stenothyridae - about 60 freshwater species, others are marine.

Neogastropoda

- Buccinidae - 8-10 freshwater species in the genus *Clea*, native to Southeast Asia. Other Buccinidae are marine.

- Marginellidae - 2 freshwater species in the genus *Rivomarginella*, native to Southeast Asia. Other Marginellidae are marine.

Heterobranchia

Family Valvatidae, shells of *Valvata sibirica*, scale is in mm

Acochlidium fijiiensis is one of very few freshwater gastropods without a shell.

Lower Heterobranchia

- Glacidorbidae - 20 species.

- Valvatidae, small low-spired snails referred to as "valve snails". 71 species.

Acochlidiacea

- Acochlidiidae (including synonym Strubelliidae) - 5 shell-less species: *Acochlidium amboinense, Acochlidium bayerfehlmanni, Acochlidium fijiiensis, Palliohedyle sutteri* and *Strubellia paradoxa*

- Tantulidae - there is only one species which is shell-less *Tantulum elegans.*

Pulmonata, Basommatophora

Basommatophorans are pulmonate or air-breathing aquatic snails, characterized by having their eyes located at the base of their tentacles, rather than at the tips, as in the true land snails Stylommatophora. The majority of basommatophorans have shells that are thin, translucent, and relatively colorless, and all five freshwater basommatophoran families lack an operculum.

- Chilinidae, small to medium-sized snails confined to temperate and cold South America. About 15 species.

- Latiidae, small limpet-like snails confined to New Zealand. One or three species.

- Acroloxidae - about 40 species.

- Lymnaeidae, found worldwide, but are most numerous in temperate and northern regions. These are the dextral (right-handed) pond snails. About 100 species.

- Planorbidae, "rams horn" snails, with a worldwide distribution. About 250 species.

- Physidae, left-handed (sinistral) "pouch snails", native to Europe, Asia, North America. About 80 species.

Family Lymnaeidae, *Lymnaea stagnalis.*

Family Physidae, *Physella acuta.*

As Human Food

Several different freshwater snail species are eaten in Asian cuisine.

Archaeological investigations in Guatemala have revealed that the diet of the Maya of the Classic Period (AD 250-900) included freshwater snails.

A dish of cooked saltwater nerites from the Rajang River, Sarawak, Malaysia

A dish of cooked freshwater snails, ampullariids and viviparids from Poipet, Cambodia

Aquarium Snails

In the developed world, people encounter freshwater snails most commonly in aquaria along with tropical fish. Species available vary in different parts of the world. In the United States, commonly available species include ramshorn snails such as *Planorbella duryi*, apple snails such as *Pomacea bridgesii*, the high-spired thiarid malaysian trumpet snail *Melanoides tuberculata*, and several *Neritina* species.

Parasitology

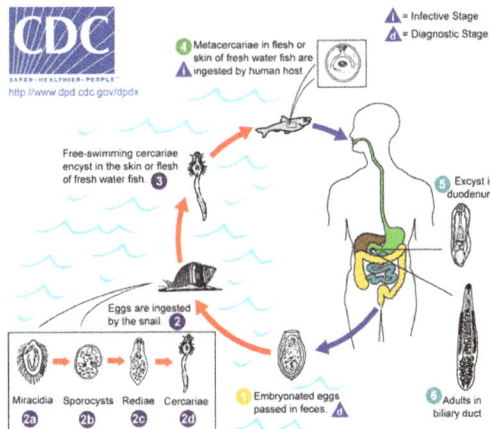

Life cycle of two liver fluke species which have freshwater snails as intermediate hosts

Freshwater snails are widely known to be hosts in the lifecycles of a variety of human and animal parasites, particularly trematodes or "flukes". Some of these relations for prosobranch snails include *Oncomelania* in the family Pomatiopsidae as hosts of *Schistosoma*, and *Bithynia*, *Parafossarulus* and *Amnicola* as hosts of *Opisthorchis*. *Thiara* and *Semisulcospira* may host *Paragonimus*. *Juga plicifera* may host *Nanophy-*

etus salmincola. Basommatophoran snails are even more widely infected, with many *Biomphalaria* (Planorbidae) serving as hosts for *Schistosoma mansoni, Fasciolopsis* and other parasitic groups. The tiny *Bulinus* snails are hosts for *Schistosoma haematobium.* Lymnaeid snails (Lymnaeidae) serve as hosts for *Fasciola* and the cerceriae causing swimmer's itch. The term "neglected tropical diseases" applies to all snail-borne infections, including schistosomiasis, fascioliasis, fasciolopsiasis, paragonimiasis, opisthorchiasis, clonorchiasis, and angiostrongyliasis.

Melanopsidae

Melanopsidae, common name melanopsids, is a family of freshwater gastropods in the clade Sorbeoconcha. Species in this family are native to southern and eastern Europe, northern Africa, parts of the Middle East, New Zealand, and freshwater streams of some large South Pacific islands.

These snails first appeared in the Late Cretaceous and are closely related to Potamididae. As well as unidirectional evolutionary change from one species to the next over time, the process of hybridization plays a major role in the appearance of new Melanopsidae species.

According to the taxonomy of the Gastropoda by Bouchet & Rocroi (2005) the family Melanopsidae has no subfamilies.

Genera

Genera in the family Melanopsidae include:

- *Amphimelania* P. Fischer, 1885

- *Esperiana* Bourguignat, 1877

- *Fagotia* Bourguignat, 1884

- *Holandriana* Bourguignat, 1884

  o *Holandriana holandrii* (C. Pfeiffer, 1828) or *Amphimelania holandrii* (C. Pfeiffer, 1828)

- *Melanopsis* Férussac, 1807 - type genus of the family Melanopsidae

- *Pseudobellardia* Cox, 1931

- *Stomatopsis* Stache, 1871

- *Stylospirula* Rovereto, 1899

- *Zemelanopsis* Finlay, 1927

  o *Zemelanopsis trifasciata* (Gray, 1843)

Paludomidae

Paludomidae, common name paludomids, is a family of freshwater snails, gastropod molluscs in the clade Sorbeoconcha.

Distribution

The distribution of the Paludomidae includes Asia and Africa.

Taxonomy

The following three subfamilies have been recognized in the taxonomy of Bouchet & Rocroi (2005):

- Paludominae Stoliczka, 1868 - synonym: Philopotamidinae Stache, 1889

- Cleopatrinae Pilsbry & Bequaert, 1927

- Hauttecoeuriinae Bourguignat, 1885

  o tribe Hauttecoeuriini Bourguignat, 1885 - synonym: Tanganyiciinae Bandel, 1998

  o tribe Nassopsini Kesteven, 1903 - synonym: Lavigeriidae Thile, 1925

  o tribe Rumellini Ancey, 1906

  o tribe Spekiini Ancey, 1906 - synonyms: Giraudiidae Bourguignat, 1885 (inv.); Reymondiinae Bandel, 1998

  o tribe Syrnolopsini Bourguignat, 1890

  o tribe Tiphobiini Bourguignat, 1886 - synonyms: Hilacanthidae Bourguignat, 1890; Paramelaniidae J. E. S. Moore, 1898; Bathanaliidae Ancey, 1906; Limnotrochidae Ancey, 1906

Genera

Genera within the family Paludomidae include:

Paludominae

- *Paludomus* Swainson, 1840 - type genus of the family Paludomidae

Cleopatrinae

- *Cleopatra* Troschel, 1857 - type genus of the subfamily Cleopatrinae

Hauttecoeuriinae

tribe Hauttecoeuriini

- *Tanganyicia* Crosse, 1881 - synonym: *Hauttecoeuria* Bourguignat, 1885

tribe Nassopsini

- *Lavigeria* Bourguignat, 1888

- *Nassopsis* E. A. Smith, 1890 - type genus of the tribe Nassopsini

tribe Rumellini

- *Rumella* Bourguignat, 1885 - type genus of the tribe Rumellini

tribe Spekiini

- *Bridouxia* Bourguignat, 1885

- *Reymondia* Bourguignat, 1885

- *Spekia* Bourguignat, 1879 - type genus of the tribe Spekiini

tribe Syrnolopsini

- *Syrnolopsis* E. A. Smith, 1880 - type genus of the tribe Syrnolopsini

tribe Tiphobiini

- *Bathanalia* Moore, 1898

- *Chytra* Moore, 1898 - with the only species *Chytra kirki* (E. A. Smith, 1880)

- *Limnotrochus* Smith, 1880 - with the only species *Limnotrochus thomsoni* Smith, 1880

- *Mysorelloides* Leloup, 1953 - with the only species *Mysorelloides multisulcata* (Bourguignat, 1888)

- *Paramelania* Smith, 1881

- *Tiphobia* E. A. Smith, 1880 - type genus of the tribe Tiphobiini, with the only species *Tiphobia horei* E. A. Smith, 1880

subfamily ?

- *Anceya* Bourguignat, 1885

- *Hirthia* Ancey, 1898

- *Martelia* Dautzenberg, 1908

- *Potadomoides* Leloup, 1953

- *Pseudocleopatra* Thiele, 1928

- *Stanleya* Bourguignat, 1885 - probably with the only species *Stanleya neritinoides* Smith, 1880

- *Stormsia* Leloup, 1953 - with the only species *Stormsia minima* (Smith, 1908)

- *Vinundu* Michel, 2004

Freshwater Fish

Tench are common freshwater fish throughout temperate Eurasia.

Freshwater fish are those that spend some or all of their lives in fresh water, such as rivers and lakes, with a salinity of less than 0.05%. These environments differ from marine conditions in many ways, the most obvious being the difference in levels of salinity. To survive fresh water, the fish need a range of physiological adaptations.

41.24% of all known species of fish are found in fresh water. This is primarily due to the rapid speciation that the scattered habitats make possible. When dealing with ponds and lakes, one might use the same basic models of speciation as when studying island biogeography.

Physiology

Freshwater fish differ physiologically from salt water fish in several respects. Their gills must be able to diffuse dissolved gasses while keeping the salts in the body fluids inside. Their scales reduce water diffusion through the skin: freshwater fish that have lost too many scales will die. They also have well developed kidneys to reclaim salts from body fluids before excretion.

Migrating Fish

Many species of fish do reproduce in freshwater, but spend most of their adult lives in the sea. These are known as anadromous fish, and include, for instance, salmon, trout and three-spined stickleback. Some other kinds of fish are, on the contrary, born in salt water, but live most of or parts of their adult lives in fresh water; for instance the eels. These are known as catadromous fish.

Sturgeons are found both in anadromous and fresh water stationary forms

Species migrating between marine and fresh waters need adaptations for both environments; when in salt water they need to keep the bodily salt concentration on a level lower than the surroundings, and vice versa. Many species solve this problem by associating different habitats with different stages of life. Both eels, anadromous salmoniform fish and the sea lamprey have different tolerances in salinity in different stages of their lives.

Classification in the United States

Among fishers in the United States, freshwater fish species are usually classified by the water temperature in which they survive. The water temperature affects the amount of oxygen available as cold water contains more oxygen than warm water.

Coldwater

Coldwater fish species survive in the coldest temperatures, preferring a water temperature of 50 to 60 °F (10–16 °C). In North America, air temperatures that result in sufficiently cold water temperatures are found in the northern United States, Canada, and in the southern United States at high elevation. Common coldwater fish include brook trout, rainbow trout, and brown trout.

Warmwater

Warmwater fish species can survive in a wide range of conditions, preferring a water temperature around 80 °F (27 °C). Warmwater fish can survive cold winter temperatures in northern climates, but thrive in warmer water. Common warmwater fish include largemouth bass, bluegill, catfish, and crappies.

Coolwater

Coolwater fish species prefer water temperature between the coldwater and warmwater species, around 60 to 80 °F (16–27 °C). They are found throughout North America except for the southern portions of the United States. Common coolwater species include muskellunge, northern pike, walleye, and yellow perch

Status

North America

About four in ten North American freshwater fish are endangered, according to a pan-North American study, the main cause being human pollution. The number of fish species and subspecies to become endangered has risen from 40 to 61, since 1989.

Cavefish

Phreatichthys andruzzii showing the pale colour and lack of eyes typical of cavefish

Cavefish or cave fish is a generic term for fresh and brackish water fish adapted to life in caves and other underground habitats. Related terms are subterranean fish, troglomorphic fish and hypogean fish.

Being aquatic, they are a part of the troglobite group known as stygofauna. The more than 200 scientifically described species of obligate cavefish are found on all continents, except Antarctica. Although widespread as a group, many cavefish species have very small ranges and are seriously threatened. Cavefish are members of a wide range of families and do not form a monophyletic group. Typical adaptations found in cavefish are reduced eyes and pigmentation.

Adaptations

Many aboveground fish may enter caves on occasion, but obligate cavefish (fish that require underground habitats) are extremophiles with a number of unusual adaptations known as troglomorphism. In some species, notably the Mexican tetra, shortfin molly, Oman garra, *Indoreonectes evezardi* and a few catfish, both "normal" aboveground and cavefish forms exist. Many adaptions seen in cavefish are aimed at surviving in a habitat with little food. Living in darkness, pigmentation and eyes are useless, or an actual disadvantage because of their energy requirements, and therefore typically reduced in cavefish. Other examples of adaptations are larger fins for more energy-efficient swimming, and a loss of scales, swim bladder and behaviors such as certain types of display. The loss can be complete or only partial, for example resulting in small (but still existing) eyes. In some cases, "blind" cavefish may still be able to see; juvenile Mexican tetras of the cave form are able to sense light via certain cells in the pineal gland (pineal eye). In the most extreme cases, the lack of light has changed the circadian rhythm (24-hour internal body clock) of the cavefish. In the Mexican tetra of the cave form and in *Phreatichthys andruzzii* the circadian rhythm lasts 30 hours and 47 hours, respectively. This may help them to save energy. Without sight, other senses are used and these may be enhanced. Examples include the lateral line for sensing vibrations, chemoreception (via smell and taste buds), electroreception, and mouth suction to sense nearby obstacles (comparable to echolocation). The level of specialized adaptations in a cavefish is generally considered to be directly correlated to the amount of time it has been restricted to the underground habitat, with species that recently arrived showing few adaptations and species with the largest number of adaptations likely being the ones restricted to the habitat for the longest time.

As typical of cavefish, *Typhleotris madagascariensis* is an
opportunistic feeder on various invertebrates

Cavefish are quite small with most species being less than 10 cm (4 in) long and very few able to surpass 20 cm (8 in). At up to about 40 cm (16 in), the blind cave eel is the longest known cavefish. The very limited food resources in the habitat likely prevents larger cavefish species from existing and also means that cavefish in general are opportunistic feeders, taking whatever is available. In their habitat, cavefish are often the top predators, feeding on smaller cave-living invertebrates, or are detritivores without enemies. Cavefish typically have low metabolic rates and may be able to survive long periods of starvation. A captive *Phreatobius cisternarum* did not feed for a year, but remained in good condition. The cave form of the Mexican tetra can build up unusually large fat reserves by "binge eating" in periods where food is available, which then (together with its low metabolic rate) allows it to survive without food for months, much longer than the aboveground form of the species.

Some fish species that live buried in the bottom of aboveground waters, live deep in the sea or in deep rivers have adaptations similar to cavefish, including reduced eyes and pigmentation.

Habitat

The Mexican blind brotula and other cave-dwelling brotulas are among the few species that live in anchialine habitas

Although many cavefish species are restricted to underground lakes, pools or rivers in actual caves, some are found in aquifers and may only be detected by humans when artificial wells are dug into this layer. Most live in areas with low (essentially static) or moderate water current, but there are also species in places with very strong current, such as the waterfall climbing cave fish. Underground waters are often very stable environments with limited variations in temperature (typically near the annual average of the surrounding region), nutrient levels and other factors. Organic compounds generally only occur in low levels and rely on outside sources, such as contained in water that enters the underground habitat from outside, aboveground animals that find their way into caves (deliberately or by mistake) and guano from bats that roost in caves. Cavefish are primarily restricted to freshwater. A few species, notably the cave-dwelling brotulas, *Luciogobius* gobies, *Milyeringa* sleeper gobies and the blind cave eel, live in anchialine caves and several of these tolerate various salinities.

Range and Diversity

The more than 200 scientifically described obligate cavefish species are found in most continents, but there are strong geographic patterns and the species richness varies. The vast majority of species are found in the tropics or subtropics. Cavefish are strongly linked to regions with karst, which commonly result in underground sinkholes and subterranean rivers.

With more than 120 described species, by far the greatest diversity is in Asia, followed by more than 30 species in South America and about 30 species in North America. In contrast, only 9 species are known from Africa, 5 from Oceania, and 1 from Europe. On a country level, China has the greatest diversity with more than 80 species, followed by Brazil with more than 20 species. India, Mexico, Thailand and the United States of America each have 9–12 species. No other country has more than 4 cavefish species.

The Hoosier cavefish from Indiana in the United States was only described in 2014

Being underground, many places where cavefish may live have not been thoroughly surveyed. New cavefish species are described with some regularity and undescribed species are known. As a consequence, the number of known cavefish species has risen rapidly in recent decades. In the early 1990s only about 50 species were known, in 2010 about 170 species were known, and by 2015 this had surpassed 200 species. It has been estimated that the final number might be around 250 obligate cavefish species. For example, the first cavefish in Europe, a *Barbatula* stone loach, was only discovered in 2015 in Southern Germany. Conversely, their unusual appearance means that some cavefish already attracted attention in ancient times. The oldest known description of an obligate cavefish, involving *Sinocyclocheilus hyalinus*, is almost 500 years old.

Obligate cavefish are known from a wide range of families: Characidae (characids), Balitoridae (hillstream loaches), Cobitidae (true loaches), Cyprinidae (carps and allies), Nemacheilidae (stone loaches), Amblycipitidae (torrent catfishes), Astroblepidae (naked sucker-mouth catfishes), Callichthyidae (armored catfishes), Clariidae (airbreathing catfishes), Heptapteridae (heptapterid catfishes), Ictaluridae (ictalurid catfishes), Kryptoglanidae (kryptoglanid catfish), Loricariidae (loricariid catfishes), Phreatobiidae (phreatobiid catfishes), Trichomycteridae (pencil catfishes), Sternopygidae (glass knifefishes), Amblyopsidae (U.S. cavefishes), Bythitidae (brotulas), Poeciliidae (live-bearers), Synbranchidae (swamp eels), Cottidae (true sculpins), Eleotridae (sleeper gobies) and Gobiidae (gobies). Many of these families are only very distantly related and do not form a monophyletic group, showing that adaptations to a life in caves has happened numerous times among fish. As such, the descriptive term "cavefish" is an example of folk taxonomy rather than scientific taxonomy. Strictly speaking some Cyprinodontidae (pupfish) are also known from sinkhole caves, famously including the Devils Hole pupfish, but these lack the adaptations (e.g., reduced eyes and pigmentation) typically associated with cavefish. Additionally, species from a few families such as Chaudhuriidae (earthworm eels), Glanapteryginae and Sarcoglanidinae live buried in the bottom of aboveground waters, and can show adaptions similar to traditional underground-living (troglobitic) fish. It has been argued that such species should be recognized as a part of the group of troglobitic fish.

Species

As of 2017, the following underground-living fish species with various levels of troglo-morphism (ranging from complete loss of eyes and pigment, to only a partial reduction of one of these) are known. *Prietella phreatophila*, the only species with underground populations in more than one country, is listed twice. Excluded from the table are species that live burried in the bottom of aboveground waters (even if they have troglomorphic-like features) and undescribed species.

Conservation

The cave form of the Mexican tetra is easily bred in captivity and the only cavefish widely available to aquarists

Although cavefish as a group are found throughout large parts of the world, many cavefish species have tiny ranges (often restricted to a single cave or cave system) and are seriously threatened. For example, the Alabama cavefish is only found in the Key Cave and the entire population has been estimated at less than 100 individuals, while the golden cave catfish only is found in the Aigamas cave in Namibia and has an estimated population of less than 400 individuals. The Haditha cavefish from Iraq and the Oaxaca cave sleeper from Mexico may already be extinct, as recent surveys have failed to find them. In some other cases, such as the Brazilian blind characid which went unrecorded by ichthyologists from 1962 to 2004, the apparent "rarity" was likely because of a lack of surveys in its range and habitat, as locals considered it relatively common until the early 1990s (more recently, this species appears to truly have declined significantly). Living in very stable environments, cavefish are likely more vulnerable to changes in the water (for example, temperature or oxygen) than fish of aboveground habitats which naturally experience greater variations. The main threats to cavefish are typically changes in the water level (mainly through water extraction or drought), habitat degradation and pollution, but in some cases introduced species and collection for the aquarium trade also present a threat. Cavefish often show little fear of humans and can sometimes be caught with the bare hands. Most cavefish lack natural predators, although larger cavefish may feed on smaller individuals, and cave-living crayfish, crabs, giant water bugs and spiders have been recorded feeding on a few species of cavefish.

Caves in some parts of the world have been protected, which can safeguard the cavefish. In a few cases such as the Omani blind cave fish (Oman garra), zoos have initiated breeding programs as a safeguard. In contrast to the rarer species, the cave form of the Mexican tetra is easily bred in captivity and widely available to aquarists. This is the most studied cavefish species and likely also the most studied cave organism overall. As of 2006, only six other cavefish species have been bred in captivity, typically by scientists.

Freshwater Drum

The freshwater drum, *Aplodinotus grunniens*, is a fish endemic to North and Central America. It is the only species in the genus *Aplodinotus*. The freshwater drum is a member of the family Sciaenidae, and is the only North American member of the group that inhabits freshwater for its entire life. Its generic name, *Aplodinotus*, comes from Greek meaning "single back", and the specific epithet, *grunniens*, comes from a Latin word meaning "grunting". It is given to it because of the grunting noise that mature males make. This noise comes from a special set of muscles within the body cavity that vibrate against the swim bladder. The purpose of the grunting is unknown, but due to it being present in only mature males, it is assumed to be linked to spawning.

The drum typically weighs 5–15 lb (2.3–6.8 kg). The world record was caught on Nickajack Lake in Tennessee, and weighed in at 54 lb 8 oz (24.7 kg). The freshwater drum is gray or silvery in turbid waters and more bronze or brown colored in clearer waters. It is a deep bodied fish with a divided dorsal fin consisting of 10 spines and 29–32 rays. It is also called shepherd's pie, gray bass, Gasper goo, Gaspergou, gou, grunt, grunter, grinder, wuss fish, Gooble Gobble, and croaker, and is commonly known as sheephead or sheepshead in parts of Canada, the United Kingdom, and the United States.

Geographic Distribution

Freshwater drum are the only North American member of their family to exclusively inhabit freshwater (freshwater family members in genera *Pachyurus* and *Plagioscion* are from South America, while *Boesemania* is Asian). Their great distribution range goes as far north as the Hudson Bay, and reaches as far south as Guatemala. Their longitudinal distribution goes as far east as the eastern Appalachians and stretches as far west into Texas, Kansas, and Oklahoma. Freshwater drum are considered to be one of the most wide-ranging species in North America.

Ecology

The freshwater drum prefers clear water, but it is tolerant of turbid and murky water. They prefer the bottom to be clean sand and gravel substrates.

The diet of the freshwater drum is generally benthic and composed of macroinvertebrates (mainly aquatic insect larvae and bivalve mussels), as well as small fish in certain ecosystems. Freshwater drum show distinct seasonal differences in their diet. In April and May, the drum feeds on dipterans. During these months, dipterans make up about 50 percent of the freshwater drum's diet. In August through November, they tend to eat fish (which are primarily young-of-the-year gizzard shad). The percentage of fish in their diet at this time ranges from 52-94 percent. Other items in the drum's diet are mollusks and crayfish. Freshwater drum tend to hang out with walleye.

The freshwater drum competes with several organisms. During its early stages in Lake Erie, it has been shown to compete with yellow perch, the trout-perch, and the emerald shiner. During its adult lifetime, it competes with yellow perch and silver chub in deep water, and competes with black bass in the shoal areas.

Predators on drum include humans and other fish. During its first year, the freshwater drum serves as a forage fish for many species of predatory fish. These include small-mouth bass, walleye, and many other piscivores. After its first year, the primary predators on freshwater drum are humans. The drum is an important commercial crop on the Mississippi River, but in other areas it constitutes only a small portion of the commercial catch. Consistent with other Sciaenids, freshwater drum are strongly nocturnal with the bulk of most catches being derived from night angling/sampling. Commercial fisheries are present for this species, although market price tends to be quite low. Thus, many freshwater drum are harvested as bycatch from targeted higher-value species.

There has been some research on the freshwater drum's impact on the invasive Zebra mussel in northern lakes and rivers. Zebra mussels are consumed by freshwater drum once they reach a length of 25 cm (9.8 in), but drum under 35 cm (14 in) in length only eat small mussels and reject the larger ones. The fish larger than 35 cm (14 in) exhibit less selectivity and consume mussels relative to their availability in lakes. These larger fish are not restricted by their ability to crush the zebra mussels, but they are restricted by the size of the clumps that they can remove. Though the drum do eat zebra mussels, they are not having an impact on the spread of this invasive species. Though they do not control the population of zebra mussels, they do contribute to a high mortality in the zebra mussels.

Life History

During the summer, freshwater drum move into warm, shallow water that is less than 33 ft (10 m) deep. The freshwater drum then spawn during a six to seven-week period from June through July when the water reaches a temperature of about 65 °F (18 °C). During the spawn, females release their eggs into the water column and males release their sperm. Fertilization is random. Males generally reach sexual maturity at four years, whereas females reach maturity at five or six years. Females from six to nine years old have a clutch size of 34,000 to 66,500 eggs and they spawn in open water giving no parental care to their larvae. The eggs then float to the top of the water column and hatch between two and four days. Due to the broadcasting of eggs in open water and lack of parental care, many eggs and larvae fall victim to predation upon hatching, the pro-larvae average 3.2 mm (0.13 in) long. The post larval stage begins about 45 hours after hatching and a length of 4.4 mm (0.17 in) is attained.

Typical freshwater drum, Lake Jordan, Alabama (released)

Females grow at a faster rate than the males and adult characteristics start to form at a length of 15 mm (0.59 in). Females continue to outgrow the male throughout their lives reaching a length of 12 to 30 in (30 to 76 cm). Usually the freshwater drum weighs 2–10 lb (0.91–4.54 kg), but they can reach well over 36 lb (16 kg). Freshwater drum are long-lived and have attained maximum ages of 72 years old in Red Lakes, Minnesota and 32 years old in the Cahaba River, Alabama. Using sectioned otoliths from archaeological sites near Lake Winnebago, Wisconsin, freshwater drum have attained the age 74 years. Though they can reach a very old age, the average age of a freshwater drum is between 6 and 13 years.

Current Management

There are not currently any management practices for the species. The freshwater drum is not federally or state listed by any states. Although the commercial harvest is up to 1 million pounds per year, they are in no danger of overharvest. In the Mississippi River alone, the commercial catch has reached about 300,000 lb (140,000 kg) in recent years. Due to its abundance, many states allow bowfishing and other non-conventional means to harvest the fish.

Australian Bass

Australian bass (*Macquaria novemaculeata*) are a small to medium-sized, primarily freshwater (but estuarine spawning) native fish found in coastal rivers and streams along the east coast of Australia. They are a member of the Percichthyidae family and *Macquaria* genus (although some recognize *Percalates* instead). Australian bass are an iconic, highly predatory native fish. They are an important member of the native fish faunas found in east coast river systems and an extremely popular angling species. The species was simply called perch in most coastal rivers where it was caught until the 1960s, when the name Australian bass started to gain popularity.

Taxonomy

An Australian bass (summer, freshwater reaches) before release.

Australian bass (*Macquaria novemaculeata*) are closely related and very similar in appearance to estuary perch (*Macquaria colonorum*). Estuary perch however tend to remain in the estuarine reaches or (occasionally) the extreme lower freshwater reaches.

Until very recently (i.e. late 2000s), Australian bass and estuary perch were placed in the *Macquaria* genus — one of a number of Australian genera in the Percichthyidae family — along with two species of native perch from the Murray-Darling Basin, golden perch (*Macquaria ambigua*) and Macquarie perch (*Macquaria australasica*). This revision to their taxonomy occurred in the late 1970s. Prior to that, Australian bass and estuary perch were placed in a separate genus, *Percalates*. (Interestingly, the generic name *Percalates* is a compound of the generic names *Perca* and *Lates*, and arose from an early, erroneous taxonomic belief that Australian bass were an old world perch related to barramundi (*Lates calcarifer*)).

Results from recent research using genetic MtDNA analysis indicate Australian bass and estuary perch do belong in a separate genus to golden perch and Macquarie perch, and has resulted in Australian bass and estuary perch being placed back into a resurrected *Percalates* genus.

A rather surprising and unexpected finding of this research is that the *Percalates* genus (i.e. Australian bass and estuary perch) appears to be genetically closer to the *Maccullochella* genus (i.e. Murray cod and other cod species) than the remnant *Macquaria* genus is (i.e. golden perch and Macquarie perch).

The specific name for Australian bass — *novemaculeata* — coined by Steindachner when he scientifically described and named the species, translates literally from the Latin as "new" (*novem*) and "spotted" (*maculeata*), and refers to the distinct black blotches juvenile bass are marked with when very small (i.e. <6–7 cm).

Description and Size

Australian bass have a moderately deep, elongated body that is laterally compressed. They have a forked caudal ("tail") fin and angular anal and soft dorsal fins. Their spiny dorsal fin is of medium height, strong and sharp. They have a medium-sized mouth and relatively large eyes than can appear dark in low light or red in bright light. The opercula or gill covers on Australian bass carry extremely sharp flat spines that can cut fishermens' fingers deeply.

Australian bass vary in colour from xanthic (a shade of yellow), or gold in clear sandy streams to the more usual bronze or bronze-green colouration in streams with darker substrates and/or some tannin staining to the water.

Australian bass are, overall, a smallish-sized species, averaging in most waters around 0.4–0.5 kg and 20–30 cm. A fish of 1 kg or larger is a good specimen. Maximum size appears to be around 2.5 kg and 55 cm in southern waters, and around 3.0 kg and 60–65 cm in northern waters.

Typically, Australian bass stocked in man-made impoundments (where they cannot breed) show greater average and maximum sizes than wild river fish.

Range

Australian bass are found in coastal rivers and streams from Wilsons Promontory in Victoria east and north along the eastern seaboard to the rivers and creeks of the Bundaberg region in central Queensland.

Australian bass are not found in the Murray-Darling system because although the system is extensive, it has only one variable entrance to the Southern Ocean, a feature that appears to be incompatible with the estuarine breeding habits of Australian bass and other aspects of their life cycle.

Migratory Patterns

Australian bass are primarily a freshwater riverine species, but must breed in estuarine waters. Consequently, Australian bass reside in the freshwater reaches of coastal rivers for the warmer half of the year or slightly more and the estuarine reaches in winter, and are highly migratory in general.

A general description of the migratory pattern for adult Australian bass would be:

- September: re-enter lower freshwater reaches after spawning

- October–November: movement through middle freshwater reaches

- December–February: maximum penetration into negotiable upper freshwater reaches

- March–April: slow movement back down through freshwater reaches in anticipation of spawning run

- May: strong spawning run to estuarine reaches

- June–July–August: aggregation and spawning in estuarine reaches

Obviously the timing of these migratory movements varies slightly from the south to the north of their range. The timing of these migratory movements are also dependent on river flows, particularly freshes and floods that drown out and make larger rapids and cascades passable.

A large wild female Australian bass (early autumn, freshwater reaches)
before release. This specimen was making her way down to the estuary for winter spawning.

Australian bass are found at their highest altitude in the freshwater reaches of rivers during the months of December, January and February. Research indicates there is sexual segregation in this non-spawning season for resource partitioning purposes. Males inhabit the lower freshwater reaches of rivers while females travel far into the middle and even upper freshwater (upland) reaches. The distance Australia bass travel upstream appears to be limited only by flows and impassable barriers (historically, waterfalls; today, often, dams). Thus, historically, the effective altitudinal limit for Australian bass has been as high as 400–600 metres in some river systems. For instance, Australian bass originally migrated up to the Dalgety region in the Snowy River, well above Oallen Crossing on the Shoalhaven River and far up the Warragamba River and Coxs River before these rivers were dammed:

> *"Messrs R. F. Seymour and L. Whitfeld were also out for perch, and the first-named did some exploring along the valley of the Cox River and the head of the Burragorang Valley. Mr Seymour found the perch plentiful in the Cox, but his travelling proved very rough." 'Angling', SMH, 8 October 1910.*

Habitat

In the freshwater reaches of coastal rivers in the warmer months, Australian bass require reasonable quality, unsilted habitats with adequate native riparian vegetation and in-stream cover/habitat. Australian bass generally sit in cover during the day. However, they are fairly flexible about the type of cover used. Sunken timber ("snags"), undercut banks, boulders, shade under trees and bushes overhanging the water and thick weedbeds are all used as cover. Such cover does not need to be in deep water to be used; Australian bass are happy to use cover in water as shallow as 1 metre in depth.

Australian bass are strong swimmers at all sizes and can easily traverse rapids and fast-flowing water. However, they generally avoid sitting directly in currents to conserve energy.

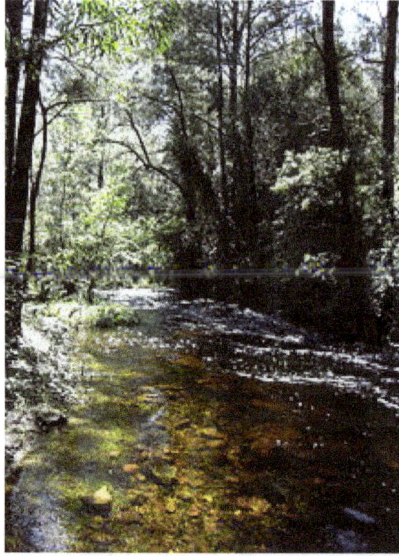

Australian bass easily traverse rapids like these in their coastal river habitats. However, they require floods or freshes to drown out more significant rapids and cascades and make them passable.

At night Australian bass display pelagic ("near-surface") behaviour and actively hunt prey in shallow water and at the water's surface.

When aggregated for spawning in the broad reaches of estuaries in winter, Australian bass are less cover oriented, and generally sit in deeper water.

Diet

Common items in the diet of Australian bass are:

- terrestrial insects, particularly cicadas

- aquatic macroinvertebrates, particularly Trichoptera larvae

- crustaceans in the forms of freshwater shrimps and estuarine prawns

- small fish, particularly flathead gudgeon (*Philhypnodon grandiceps*), which are common in their freshwater habitats.

However, Australian bass are fierce predators and any small creature that swims across a bass pool such as (introduced) mice and native lizards or frogs are at risk of being taken by a large Australian bass, and are regularly taken.

Growth and Age

For reasons that are not clear, Australian bass are extremely slow growing. Australian bass continue the trend present in the larger native fish species of SE Australia of being very long-lived. Longevity is a survival strategy to ensure that most adults participate in at least one exceptional spawning and recruitment event, which are often linked to unusually wet 'La Niña' years and may only occur every one or two decades. Maximum age recorded so far is 22 years.

As with other *Macquaria* species, there is sexual dimorphism in Australian bass. Males tend to have an absolute maximum size of 1.0 kg or less, while females regularly exceed 1.0 kg and sometimes reach the maximum size of 2.5–3.0 kg. Males reach sexual maturity at around 3–4 years of age, females at 5–6 years of age.

Reproduction

Australian bass spawn in estuaries in winter, generally in the months of July or August.

The salinity range in which Australian bass spawn is still not clear. Estuaries are dynamic habitats with daily fluxes in salinity due to tides, and are also affected by droughts, floods and freshes (minor, temporary rises in flow), making measurements of preferred spawning salinities for wild Australian bass difficult.

Australian bass spawn in salinities of 8–12 parts per thousand (salt water is approximately 36 ppt), based on capture of recently spawned larval and juvenile Australian bass in estuaries. Australian bass sperm have no viability at or below 6 ppt, but are most viable at 12 ppt, the latter probably being the most relevant fact. However, it has been reported that Australian bass spawned in salinities of 12–18 ppt, with this statement based on fishermens' reports of observing wild Australian bass spawnings and some unpublished data gathered by the NSW Fisheries Department.

Artificial breeding of Australian bass is carried out at much higher salinities than natural.

Australian bass are highly fecund, with a reported mean fecundity ("fertility") of 440,000 eggs from the mature wild female specimens examined, and one very large specimen yielding 1,400,000 eggs. The eggs are reported as being demersal ("sinking") in natural spawning salinities, in which case estuarine vegetation such as sea grass almost certainly play an important role in "trapping" and protecting eggs. Larvae hatch in 2–3 days. Juvenile Australian bass migrate into the freshwater reaches after spending several months in estuarine waters.

Despite spawning in estuaries, Australian bass rely on floods coming down river systems into the estuaries throughout the winter period, both to stimulate migration and spawning in adult Australian bass and for strong survival and recruitment of Australian bass larvae.

Australian bass adults and larvae may also enter the sea (the latter perhaps involuntarily) during winter spawning in times of flood. It has been reported:

The presence of field-caught larvae of both species on incoming tides in Swansea Channel indicates that the larvae have spent some time in the ocean... Macquaria novemaculeata *adults move downstream into estuaries to spawn in water of suitable salinity. In low rainfall years, the spawning location is further upstream than in wet years, when spawning can occur in shallow coastal waters adjacent to estuaries (Searle, pers. comm.). Mature* M. novemaculeata *adults can be found outside of estuaries in wet years (Williams, 1970). This is verified by the collection of mature adults by trawl in July 1995 in 11–17 m of water off Newcastle, NSW (AMS I.37358-001).*

This kind of movement leads to some genetic interchange between river systems and is important in maintaining a high degree of genetic homogeneity ("sameness") in Australian bass stocks and preventing speciation. However, this movement has not prevented distinct genetic profiles and subtle morphological ("body shape") differences developing in different river systems. These findings indicate it is important to use the appropriate regional Australian bass stocks for artificial breeding and stocking projects.

Conservation

Wild Australian bass stocks have declined seriously since European settlement.

Dams and weirs blocking migration of Australian bass both to estuaries and to the upper freshwater reaches of coastal rivers is the most potent cause of decline. Most coastal rivers now have dams and weirs on them. If Australian bass are prevented from migrating to estuaries for breeding by an impassable dam or weir, then they will die out above that dam or weir. Some dams or weirs exclude Australian bass from the vast majority of their habitat. It is estimated for example that Tallowa Dam on the Shoalhaven River, once an Australian bass stronghold, currently excludes wild Australian bass from more than 80% of their former habitat (in early 2010 however a "fish lift" was fitted to the dam). Dams and weirs also diminish or completely remove flood events required for effective breeding of adult bass and effective recruitment of juvenile Australian bass. A related issue is the myriad of other structures on coastal rivers such as poorly designed road crossings that (often needlessly) block migration of Australian bass.

Another potent cause of decline is habitat degradation. Unfortunately poor land management practices have been the norm historically in Australia. Complete clearing of riparian (river bank) vegetation, stock trampling river banks, and massive siltation from these poor practices as well as poor practices in the catchment, can severely degrade and silt coastal rivers to the point of being uninhabitable for Australian bass. The Bega River in southern New South Wales is a particularly salutory example of a coastal river so stripped of riparian vegetation and so silted with coarse granitic sands from poor land management practices, that the majority of it is now completely uninhabitable by Australian bass and other native fish.

As a slow-growing fish, Australian bass are vulnerable to overfishing, and overfishing has been a driver of decline in Australian bass stocks in past decades. However, the situation has improved markedly now the majority of fishermen are practicing catch and release with Australian bass.

Hatchery breeding and stocking of Australian bass is used to create fisheries above dams and weirs but these are causing concern over genetic diversity issues, use of bass broodfish from different genetic strains, and introduction/translocation of unwanted pest fish species in stockings. Stockings can also mask and divert attention away from serious habitat degradation and decline of wild stocks in catchments.

Fishing

Fishing for Australian bass is a summertime affair, undertaken during the warmer months in the freshwater reaches of the rivers they inhabit. Australian bass are keenly fished for as they are an outstanding sportsfish, extraordinarily fast and powerful for their size. Their extraordinary speed and power is probably due to their significant, strenuous annual migrations for spawning and a life-style that is migratory in general. Australian bass in their natural river habitats are not to be underestimated; they head straight for the nearest snags (sunken timber) when hooked and light but powerful tackle and stiff drag settings are needed to stop them.

This sizeable wild Australian bass was caught on a fizzing surface lure equipped with barbless hooks (summer, freshwater reaches), and was carefully released.

As mentioned above, during the day Australian bass generally remain close to or in cover (e.g. snags, overhanging trees), and small plug lures and flies cast close to such cover are used. Australian bass will strike at a very large variety of lures from diving minnow-style hard-bodied lures to slowly jigged soft plastic baits, as well as various surface lures such as poppers and surface walkers. In recent years fly fishing for Australian bass using surface flies imitating cicadas has proven to be extremely effective. At night Australian bass are a roaming pelagic feeder and surface lures (which waddle or fizz across the surface of the water) are used.

Some of the best Australian bass fishing is coastal rivers and tributaries where access is difficult. Fishing these more remote locations can be extremely rewarding both for the fishing and the scenery. Fishing the more remote bass water is therefore usually the domain of the hardened backpacking fisherman or the dedicated kayak fisherman willing to drag his kayak over numerous logs and other obstacles.

It pays for fishermen to remember that wild Australian bass are still a highly migratory when in the freshwater reaches of rivers, and can also be an extremely wary fish in these habitats, much more so than exotic trout species.

Australian bass fishermen almost exclusively practice catch and release, which is necessary for the preservation of wild Australian bass stocks. The use of barbless hooks (which can be created by crushing the barbs flat with a pair of needle-nosed pliers) is essential as Australian bass hit lures with great ferocity and are consequently almost impossible to unhook on barbed hooks. Conversely, Australian bass are swiftly and easily released if barbless hooks are used.

Responsible fishermen now avoid fishing for wild Australian bass in estuaries in winter, so that this increasingly pressured native fish can spawn in peace. In 2014 the NSW Fisheries Department announced an extended closed season for Australian bass and estuary perch, from 1 May to 31 August.

Arctic Grayling

Arctic grayling (*Thymallus arcticus*) is a species of freshwater fish in the salmon family Salmonidae. *T. arcticus* is widespread throughout the Arctic and Pacific drainages in Canada, Alaska, and Siberia, as well as the upper Missouri River drainage in Montana. In the U.S. state of Arizona, an introduced population is found in the Lee Valley and other lakes in the White Mountains. They were also stocked at Toppings Lake by the Teton Range and in various lakes in the high Uinta Mountains in Utah.

Taxonomy

The scientific name of the Arctic grayling is *Thymallus arcticus*. It was named in 1776 by German zoologist Peter Simon Pallas from specimens collected in Russia. The name of the genus *Thymallus* first given to grayling (*T. thymallus*) described in the 1758 edition of *Systema Naturae* by Swedish zoologist Carl Linnaeus originates from the faint smell of the herb thyme, which emanates from the flesh.

Description

Arctic grayling grow to a maximum recorded length of 76 cm (30 in) and a maximum recorded weight of 3.8 kg (8.4 lb). Of typical thymalline appearance, the Arctic grayling is distinguished from the similar grayling (*T. thymallus*) by the absence of dorsal and anal spines and by the presence of a larger number of soft rays in these fins. There is a dark midlateral band between the pectoral and pelvic fins, and the flanks may possess a pink iridescence. *T. a. arcticus* has been recorded as reaching an age of 18 years.

Arctic grayling caught in the Colville River of Alaska

Range

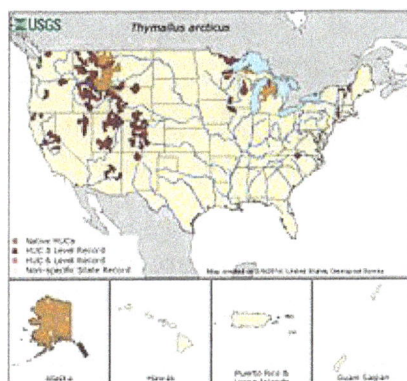
Native and introduced range of Arctic grayling, *Thymallus arcticus* in U.S.

Arctic grayling are widespread in Arctic Ocean drainages from Hudson Bay, Canada to Alaska and in Arctic and Pacific drainages to central Alberta and British Columbia in Canada. They do not occur naturally in the Fraser and Columbia river basins. There are remnant native populations of fluvial Arctic grayling in the upper Missouri River drainage in the Big Hole River and Red Rock basin ("Montana Arctic grayling"). Fluvial Arctic grayling have been reestablished in the upper Ruby River, a tributary of the Beaverhead River. The native range formerly extended south into the Great Lakes basin in Michigan. They occur naturally in the Arctic Ocean basin in Siberia from the Ob to Yenisei drainages and in European Russia in some tributaries of Pechora river. Lake dwelling forms of Arctic grayling have been introduced in suitable lake habitats throughout the Rocky Mountains including lakes in the Teton Range in Wyoming and the high Uinta Mountains in Utah), Cascade Mountains and Sierra Nevada Mountains as far south as Arizona.

Life Cycle

Several life history forms of Arctic grayling occur: fluvial populations that live and spawn in rivers; lacustrine populations that live and spawn in lakes; and potamodromous populations that live in lakes and spawn in tributary streams.

The Arctic grayling occurs primarily in cold waters of mid-sized to large rivers and lakes, returning to rocky streams to breed. The various subspecies are omnivorous. Crustaceans, insects and insect larvae, and fish eggs form the most important food items. Larger specimens of *T. arcticus* become piscivorous and the immature fish feed on zooplankton and insect larvae.

10 in (25 cm) Arctic grayling from the Gulkana River, Paxson, Alaska.

Spawning takes place in the spring. Adult fish seek shallow areas of rivers with fine, sand substrate and moderate current. Males are territorial and court females by flashing their colourful dorsal fins; the fins are also used to brace receptive females during the vibratory release of milt and roe. The fish are nonguarders: the eggs are left to mix with the substrate. Although the Arctic grayling does not excavate a nest, the highly energetic courtship and mating tends to kick up fine material which covers the zygotes. The zygote is small (approximately 3 mm or 0.1 in in diameter) and the embryo will hatch after two to three weeks. The newly hatched embryo remains in the substrate until all the yolk has been absorbed. They emerge at a length of around 12 to 18 mm (0.5 to 0.7 in), at which time they form shoals at the river margins. The juveniles grow quickly during their first two years of life.

Conservation

Arctic grayling are considered a secure species throughout their range. Although some populations at the southern extant of its native range have been extirpated, it remains widespread elsewhere and is not listed on the IUCN Red List of threatened species.

The fluvial population in the upper Missouri river basin once merited a high priority for listing under the Endangered Species Act (ESA) by the US Fish & Wildlife Service (FWS). This unique southernmost population is now extirpated from all areas of the basin with the exception of the Big Hole River watershed. In preparation for an ESA listing, the US FWS began implementing a "Candidate Conservation Agreement with Assurances" (CCAA). This agreement protects cooperating landowners from being prosecuted under the ESA "takings" clause so long as they fulfill specific obligations, spelled out in a contractual arrangement and intended to restore the dwindling population. Finally, in 2014 the FWS determined not to list the grayling under ESA, due to the effectiveness of the CCAA.

As is the grayling (*T. thymallus*), the Arctic grayling is economically important, being raised commercially for food and fished for sport.

Largetooth Sawfish

The largetooth sawfish (*Pristis microdon*), also known as the Leichhardt's sawfish or freshwater sawfish, is a sawfish of the family Pristidae, found in shallow Indo-West Pacific oceans between latitudes 11° N and 39° S. As its relatives, it also enters freshwater. This species reaches a length of up to 7 metres (23 ft). Reproduction is ovoviviparous. Recent evidence strongly suggests *P. microdon* is synonymous with *P. pristis*. Consequently, the IUCN removed *P. microdon* from their list, instead recognizing it as part of the critically endangered *P. pristis*.

Taxonomy

Considerable taxonomic confusion has surrounded this species. It is part of the *Pristis pristis* species complex, which also includes *P. perotteti*. *P. microdon* has sometimes been considered synonymous with *P. perotteti*, and uncertainty exists over what species the scientific name *P. microdon* really belong to (the original description lacked a type locality).

Recent evidence strongly suggests the three are conspecific (in which case *P. microdon* and *P. perotteti* are synonyms of *P. pristis*), as morphological and genetic differences are lacking. Three main clades based on NADH-2 genes were evident (Atlantic, Indo-West Pacific, and East Pacific), but these do not match the distributions claimed for *P. pristis* (circumtropical), *P. microdon* (Indo-West Pacific) and *P. perotteti* (Atlantic and East Pacific) respectively.

Species Description

The largetooth sawfish is a heavy-bodied sawfish with a short massive saw which is broad-based, strongly tapering and with 14 to 22 very large teeth on each side - the space between the last two saw-teeth on the sides are less than twice the space between the first two teeth. The pectoral fins are high and angular, the first dorsal fin being mostly in front of the pelvic fins, and the caudal fin has a pronounced lower lobe.

Conservation Status

The largetooth sawfish was listed as endangered in the United States under the Endangered Species Act in 2011. In December 2014 the U.S. listed the widest taxonomic definition of largetooth sawfish (including *P. perotteti* and *P. microdon*) as endangered.

Airbreathing Catfish

Airbreathing catfishes are fishes comprising the family Clariidae of order Siluriformes. About 14 genera and about 116 species of clariids are described. All the clariids are freshwater species.

Distribution

Although clariids occur in India, Syria, southern Turkey, and large parts of Southeast Asia, their diversity is the largest in Africa.

Description

Clariid catfishes are characterized by an elongated body, the presence of four barbels, long dorsal and anal fins, and especially by the autapomorphic presence of a suprabranchial organ, formed by tree-like structures from the second and fourth gill arches. This suprabranchial organ, or labyrinth organ, allows some species the capability of traveling short distances on land (walking catfishes).

The dorsal fin base is very long and is not preceded by a fin spine. The dorsal fin may or may not be continuous with the caudal fin, which is rounded. Pectoral and pelvic fins are variously absent in some species. Some fish have small eyes and reduced or absent pectoral and pelvic fins for a burrowing lifestyle. A few species are blind.

Within the Clariidae family, body forms range from fusiform (torpedo-like) to anguilliform (eel-like). As species become more eel-shaped, a whole set of morphological changes has been observed, such as decrease and loss of the adipose fin, continuous unpaired fins, reduction of paired fins, reduction of the eyes, reduction of the skull bones, and hypertrophied jaw muscles.

Taxonomy

The Heteropneustidae containing the genus *Heteropneustes* are considered by some to be a separate family and by others to be a subfamily. With the Heteropneustidae and Clariidae as separate families, a recent paper groups them into a superfamily called Clarioidea. Relationships of clarioids to other families remains uncertain.

Relationship to Humans

Many clariids form a large part of artisanal fisheries. *Clarias gariepinus* is recognized as one of the most promising aquaculture species in Africa.

The airbreathing capacity of these fish has allowed such fish as *Clarias batrachus* to be an invasive species in Florida.

Walking Catfish

The walking catfish (*Clarias batrachus*) is a species of freshwater airbreathing catfish native to Southeast Asia, but also introduced outside its native range where it is considered an invasive species. It is named for its ability to "walk" across dry land, to find food or suitable environments. While it does not truly walk as most bipeds or quadrupeds do, it has the ability to use its pectoral fins to keep it upright as it makes a wiggling motion with snakelike movements. This fish normally lives in slow-moving and often stagnant waters in ponds, swamps, streams and rivers, flooded rice paddies or temporary pools which may dry up. When this happens, its "walking" skill allows the fish to move to other sources of water. Considerable taxonomic confusion surrounds this species and it has frequently been confused with other close relatives.

Characteristics and Anatomy

The walking catfish has an elongated body shape, and reaches almost 0.5 m (1.6 ft) in length and 1.2 kg (2.6 lb) in weight. Often covered laterally in small white spots, the body is mainly coloured a gray or grayish brown. This catfish has long-based dorsal and anal fins, as well as several pairs of sensory barbels. The skin is scaleless, but covered with mucus, which protects the fish when it is out of water.

One main distinction between the walking catfish and native North American ictalurid catfish is the walking catfish's lack of an adipose fin.

This fish needs to be handled carefully when fishing it due to its hidden embedded sting or thorn-like defensive mechanism hidden behind its fins (including the middle ones before the tail fin, like the majority of all catfishes).

Taxonomy, Distribution, and Habitat

The walking catfish is a tropical species native to Southeast Asia. The native range of true *Clarias batrachus* is only confirmed from the Indonesian island of Java, but three closely related and more widespread species have frequently been confused with this species. These are *C. magur* of northeast India and Bangladesh, a likely undescribed species from Indochina, and another likely undescribed species from the Thai-Malay Peninsula, Sumatra, and Borneo. The undescribed species have both been referred to as *Clarias* aff. *batrachus*. At present, the taxonomic position of the Philippines population (called *hito* or simply "catfish" by the locals) is unclear, and it is also unclear whether South Indian populations are *C. magur* or another species. As a consequence, much information (behavioral, ecological, related to introduced populations, etc.) listed for *C. batrachus* may actually be for the closely related species that have been confused with true *C. batrachus*. True *C. batrachus*, *C. magur* and the two likely undescribed species are all kept in aquaculture.

Walking catfish thrive in stagnant, frequently hypoxic waters, and are often found in muddy ponds, canals, ditches and similar habitats. The species spends most of its time on, or right above, the bottom, with occasional trips to the surface to gulp air.

Diet

In the wild, this creature is omnivorous; it feeds on smaller fish, molluscs, and other invertebrates, as well as detritus and aquatic weeds. It is a voracious eater which consumes food rapidly, so it is particularly harmful when invasive.

As an Invasive Species

Within Asia, this species has been introduced widely. In the United States, it is established in Florida and reported in California, Connecticut, Georgia, Massachusetts, and Nevada and regarded as an invasive species.

The walking catfish was imported to Florida, reportedly from Thailand, in the early 1960s for the aquaculture trade. The first introductions apparently occurred in the mid-1960s when adult fish imported as brood stock escaped, either from a fish farm in northeastern Broward County or from a truck transporting brood fish between Dade and Broward Counties. Additional introductions in Florida, supposedly purposeful releases, were made by fish farmers in the Tampa Bay area, Hillsborough County in late 1967 or early 1968, after the state banned the importation and possession of walking catfish. Aquarium releases likely are responsible for introductions in other states. Dill and Cordone (1997) reported this species has been sold by tropical fish dealers in California for some time. They have also been spotted occasionally in the Midwest.

In Florida, walking catfish are known to have invaded aquaculture farms, entering ponds where these predators prey on fish stocks. In response, fish farmers have had to erect fences to protect ponds. Authorities have also created laws that ban possession of walking catfish.

As Food

In Thailand, this fish is known as *pla duk dan*. It is a common, inexpensive food item, prepared in a variety of ways, being often offered by street vendors, especially grilled or fried.

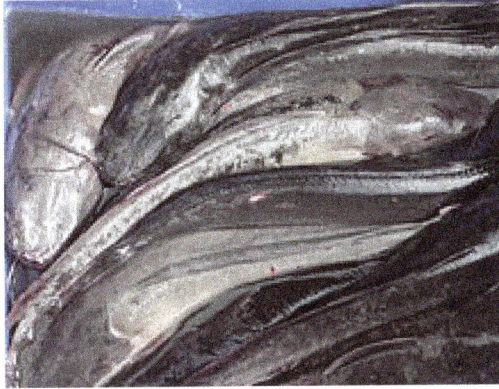

Sold in HAL market

One of the most common freshwater catfish in the Philippines, it is known as *hito* in the local language.

It is a delicacy in the Indian state of Assam, where it is called *magur machh*.

It is also eaten in West Bengal and Bangladesh.

In Indonesia, it is called *lele*, and it is the main ingredient in several native dishes, such as *pecel lele*.

In Karantaka, it is called *murgodu*. 'Murgodu' is an invasive species and is considered a threat to the native fish. Further, since this fish can stay alive even when it is out of the water, it is killed, typically, by repeatedly bashing its head using a blunt metal object. To stop this brutal practice and to protect the native breeds, several districts have banned rearing or sale of this fish. In coastal Karnataka, India, it is called *mugudu* and is considered a delicacy.

Neither the Thai nor the Indian population is likely to be the true *C. batrachus*.

Aquarium

A white variation with black patterns is commonly seen in the aquarium fish trade. However, this color variation is also prohibited where walking catfish are banned. Very well-rooted plants and large structures that provide some shade should be included. Any tankmates small enough will be eaten.

Freshwater Crocodile

The freshwater crocodile (*Crocodylus johnsoni* or *Crocodylus johnstoni*), also known as the Australian freshwater crocodile, Johnstone's crocodile or colloquially as freshie, is a species of reptile endemic to the northern regions of Australia.

Unlike their much larger Australian relative, the saltwater crocodile, freshwater crocodiles are not known as man-eaters and rarely cause fatalities, although they will bite in self-defense if cornered.

Taxonomy and Etymology

When Gerard Krefft named the species in 1873, he intended to commemorate the man who first reported it to him, Australian policeman and naturalist Robert Arthur Johnstone (1843-1905). However, Krefft made an error in writing the name, and for many years the species has been known as *C. johnsoni*. Recent studies of Krefft's papers have determined the correct spelling of the name, and much of the literature has been updated to the correct usage. However, both versions still exist. According to the rules of the International Code of Zoological Nomenclature, the epithet *johnsoni* (rather than the intended *johnstoni*) is correct.

Characteristics

The freshwater crocodile is a relatively small crocodilian. Males can grow to 2.3–3 m (7.5–9.8 ft) long, while females reach a maximum size of 2.1 m (6.9 ft). Males commonly weigh around 70 kg (150 lb), with large specimens up to 100 kg (220 lb) or more, against the female weight of 40 kg (88 lb). In areas such as Lake Argyle and Katherine Gorge there exist a handful of confirmed 4 metres (13 ft) individuals. This species is shy and has a more slender snout and have slightly smaller teeth than the dangerous saltwater crocodile. The body colour is light brown with darker bands on the body and tail—these tend to be broken up near the neck. Some individuals possess distinct bands or speckling on the snout. Body scales are relatively large, with wide, close-knit armoured plates on the back. Rounded, pebbly scales cover the flanks and outsides of the legs.

Distribution and Habitat

Freshwater crocodiles are found in the states of Western Australia, Queensland, and the Northern Territory. Main habitats include freshwater wetlands, billabongs, rivers and creeks. This species could live in areas where saltwater crocodiles could not, and are known to inhabit areas above the escarpment in Kakadu National Park and in very arid and rocky conditions (such as Katherine Gorge, where they are common and are relatively safe from saltwater crocodiles during the dry season). However, they are still consistently found in low-level billabongs, living alongside the saltwater crocodiles near the tidal reaches of rivers.

In May 2013, a freshwater crocodile was seen in a river near the desert town of Birdsville, hundreds of kilometres south of their normal range. A local ranger suggested that years of flooding may have washed the animal south, or it may have been dumped as a juvenile.

Biology and Behavior

They compete poorly with saltwater crocodiles; however, this species is saltwater tolerant. Adult crocodiles eat fish, birds, bats, reptiles and amphibians, although larger individuals may take prey as large as a wallaby.

One has even been filmed being eaten by an olive python after a struggle that lasted for an estimated 5 hours.

Reproduction

Eggs are laid in holes during the Australian dry season (usually in August) and hatch at the beginning of the wet season (November/December). The crocodiles do not defend their nests during incubation. From one to five days prior to hatching, the young begin to call from within the eggs. This induces and synchronizes hatching in siblings and stimulates adults to open the nest. It is not known if the adult that opens a given nest is the female which laid the eggs. As young emerge from the nest, the adult picks them up one by one in the tip of its mouth and transports them to the water. Adults may also assist young in breaking through the egg shell by chewing or manipulating the eggs in its mouth.

Conservation Status

Freshwater crocodile at Australia Zoo

Until recently, the freshwater crocodile was common in northern Australia, especially where saltwater crocodiles are absent (such as more arid inland areas and higher elevations). In recent years, the population has dropped dramatically due to the ingestion of the invasive cane toad. The toad is poisonous to freshwater crocodiles, although not to saltwater crocodiles, and the toad is rampant throughout the Australian wilderness. The crocodiles are also infected by *Griphobilharzia amoena*, a parasitic trematode, in regions such as Darwin.

Danger Towards Humans

Although the freshwater crocodile does not attack humans as potential prey, it can deliver a nasty bite. There have been very few incidents where people have been bitten whilst swimming with freshwater crocodiles, and others incurred during scientific study. An attack by a freshwater crocodile on a human was recorded at Barramundi Gorge (also known as Maguk) in Kakadu National Park and resulted in minor injuries; the victim managed to swim and walk away from the attack. He had apparently passed directly over the crocodile in the water. However, in general, it is still considered safe to swim with this species, so long as they are not aggravated.

Freshwater Crab

There are around 1,300 species of freshwater crabs, distributed throughout the tropics and subtropics, divided among eight families. They show direct development and maternal care of a small number of offspring, in contrast to marine crabs which release thousands of planktonic larvae. This limits the dispersal abilities of freshwater crabs, so they tend to be endemic to small areas. As a result, a large proportion are threatened with extinction.

Potamon ibericum (Potamidae) in Georgia

Parathelphusa convexa (Parathelphusidae) in Indonesia

Systematics

There are more than 1,300 described species of freshwater crabs, out of a total of 6,700 species of crabs across all environments. The total number of species of freshwater crabs, including undescribed species is thought to be up to 65% higher, potentially up to 2,155 species, although most of the additional species are currently unknown to science. They belong to eight families, each with a limited distribution, although various crabs from other families are also able to tolerate freshwater conditions (euryhaline) or are secondarily adapted to fresh water. The phylogenetic relationships between these families is still a matter of debate, and it is therefore unclear how many times the freshwater lifestyle has evolved among the true crabs. The eight families are:

Superfamily Trichodactyloidea

- Trichodactylidae (Central America and South America)

Superfamily Potamoidea

- Potamidae (Mediterranean Basin and Asia)

- Potamonautidae (Africa, including Madagascar)

- Deckeniidae (East Africa and Seychelles) – also treated as part of Potamonautidae

- Platythelphusidae (East Africa) – also treated as part of Potamonautidae

Superfamily Gecarcinucoidea

- Gecarcinucidae (Asia)

- Parathelphusidae (Asia and Australasia)

Superfamily Pseudothelphusoidea

- Pseudothelphusidae (Central America and South America)

The fossil record of freshwater organisms is typically poor, and so few fossils of freshwater crabs have been found. The oldest is *Tanzanonautes tuerkayi*, from the Oligocene of East Africa, and the evolution of freshwater crabs is likely to post-date the break-up of the supercontinent Gondwana.

Members of the family Aeglidae and *Clibanarius fonticola* are also restricted to freshwater, but these "crab-like" crustaceans are members of the infraorder Anomura (true crabs are Brachyura).

Description and Life Cycle

The external morphology of freshwater crabs varies very little, and so the form of the gonopod (first abdominal appendage, modified for insemination) is of critical importance for classification. Development of freshwater crabs is characteristically direct, where the eggs hatch as juveniles, with the larval stages passing within the egg. The broods comprise only a few hundred eggs (compared to hundreds of thousands for marine crabs) each of which is quite large, at a diameter of around 1 mm (0.04 in).

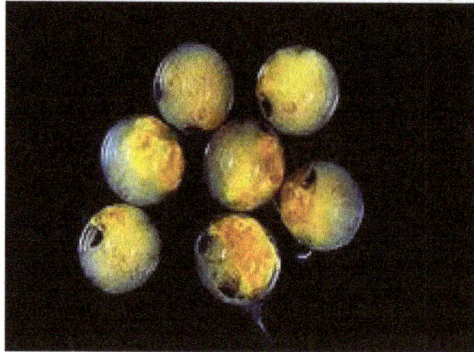

Eggs of *Potamon fluviatile* containing fully formed juvenile crabs

The colonisation of fresh water has required crabs to alter their water balance; freshwater crabs can reabsorb salt from their urine, and have various adaptations to reduce the loss of water. In addition to their gills, freshwater crabs have a "pseudo-lung" in their gill chamber that allows them to breathe in air. These developments have pre-adapted freshwater crabs for terrestrial living, although freshwater crabs need to return to water periodically in order to excrete ammonia.

Ecology and Conservation

Freshwater crabs are found throughout the tropical and sub-tropical regions of the world. They live in a wide range of water bodies, from fast-flowing rivers to swamps, as well as in tree boles or caves. They are primarily nocturnal, emerging to feed at night; most are omnivores, although a small number are specialist predators, such as *Platythelphusa armata* from Lake Tanganyika, which feeds almost entirely on snails. Some species provide important food sources for various vertebrates. A number of freshwater crabs are secondary hosts of flukes in the genus *Paragonimus*, which causes paragonimiasis in humans.

The majority of species are narrow endemics, occurring in only a small geographical area. This is at least partly attributable to their poor dispersal abilities and low fecundity, and to habitat fragmentation caused by the world's human population. In West Africa, species that live in savannahs have wider ranges than species from the rainforest; in East Africa, species from the mountains have restricted distributions, while lowland species are more widespread.

Every species of freshwater crab described so far has been assessed by the International Union for Conservation of Nature (IUCN); of the species for which data are available, 32% are threatened with extinction. For instance, all but one of Sri Lanka's 50 freshwater crab species are endemic to that country, and more than half are critically endangered.

Heron

The herons are the long-legged freshwater and coastal birds in the family Ardeidae, with 64 recognised species, some of which are referred to as "egrets" or "bitterns" rather than herons. Members of the genera *Botaurus* and *Ixobrychus* are referred to as "bitterns", and, together with the zigzag heron or zigzag bittern in the monotypic genus *Zebrilus*, form a monophyletic group within the Ardeidae. Egrets are not a biologically distinct group from the herons, and tend to be named differently because they are mainly white or have decorative plumes. Although egrets have the same build as herons, they tend to be smaller. Herons, by evolutionary adaptation, have long beaks.

The classification of the individual heron/egret species is fraught with difficulty, and no clear consensus exists about the correct placement of many species into either of the two major genera, *Ardea* and *Egretta*. Similarly, the relationships of the genera in the family are not completely resolved. However, one species formerly considered to constitute a separate monotypic family, the Cochlearidaeor the boat-billed heron, is now regarded as a member of the Ardeidae.

Although herons resemble birds in some other families, such as the storks, ibises, spoonbills, and cranes, they differ from these in flying with their necks retracted, not outstretched. They are also one of the bird groups that have powder down. Some members of this group nest colonially in trees, while others, notably the bitterns, use reed beds.

Description

The neck of this yellow bittern is fully retracted.

The herons are medium- to large-sized birds with long legs and necks. They exhibit very little sexual dimorphism in size. The smallest species is usually considered the little bittern, which can measure under 30 cm (12 in) in length, although all the species in the *Ixobrychus* genus are small and many broadly overlap in size. The largest species of heron is the goliath heron, which stands up to 152 cm (60 in) tall. The necks are able to kink in an S-shape, due to the modified shape of the cervical vertebrae, of which they have 20-21. The neck is able to retract and extend, and is retracted during flight, unlike most other long-necked birds. The neck is longer in the day herons than the night herons and bitterns. The legs are long and strong and in almost every species are unfeathered from the lower part of the tibia (the exception is the zigzag heron). In flight, the legs and feet are held backward. The feet of herons have long, thin toes, with three forward-pointing ones and one pointing backward.

The Pacific reef heron has two colour morphs, the light and the dark.

The bill is generally long and harpoon-like. It can vary from extremely fine, as in the agami heron, to thick as in the grey heron. The most atypical bill is owned by the boat-billed heron, which has a broad, thick bill. The bill, as well as other bare parts of the body, is usually yellow, black, or brown in colour, although this can vary during the breeding season. The wings are broad and long, exhibiting 10 or 11 primary feathers (the boat-billed heron has only nine), 15–20 secondaries. and 12 rectrices (10 in the bitterns). The feathers of the herons are soft and the plumage is usually blue, black, brown, grey, or white, and can often be strikingly complex. Amongst the day herons, little sexual dimorphism in plumage is seen (except in the pond-herons); differences between the sexes are the rule for the night herons and smaller bitterns. Many species also have different colour morphs. In the Pacific reef heron, both dark and light colour morphs exist, and the percentage of each morph varies geographically. White morphs only occur in areas with coral beaches.

Distribution and Habitat

Lava herons are endemic to the Galápagos Islands, where they feed on fish and crabs in the intertidal and mangrove areas.

The herons are a widespread family with a cosmopolitan distribution. They exist on all continents except Antarctica, and are present in most habitats except the coldest extremes of the Arctic, extremely high mountains, and the driest deserts. Almost all species are associated with water; they are essentially nonswimming waterbirds that feed on the margins of lakes, rivers, swamps, ponds, and the sea. They are predominantly found in lowland areas, although some species live in alpine areas, and the majority of species occurs in the tropics.

The herons are a highly mobile family, with most species being at least partially migratory. Some species are partially migratory, for example the grey heron, which is mostly sedentary in Britain, but mostly migratory in Scandinavia. Birds are particularly inclined to disperse widely after breeding, but before the annual migration, where the species is colonial, searching out new feeding areas and reducing the pressures on feeding grounds near the colony. The migration typically occurs at night, usually as individuals or in small groups.

Behaviour and Ecology

The herons and bitterns are carnivorous. The members of this family are mostly associated with wetlands and water, and feed on a variety of live aquatic prey. Their diet includes a wide variety of aquatic animals, including fish, reptiles, amphibians, crustaceans, molluscs, and aquatic insects. Individual species may be generalists or specialise in certain prey types, such as the yellow-crowned night heron, which specialises in crustaceans, particularly crabs. Many species also opportunistically take larger prey, including birds and bird eggs, rodents, and more rarely carrion. Even more rarely, herons eating acorns, peas, and grains have been reported, but most vegetable matter consumed is accidental.

Diet

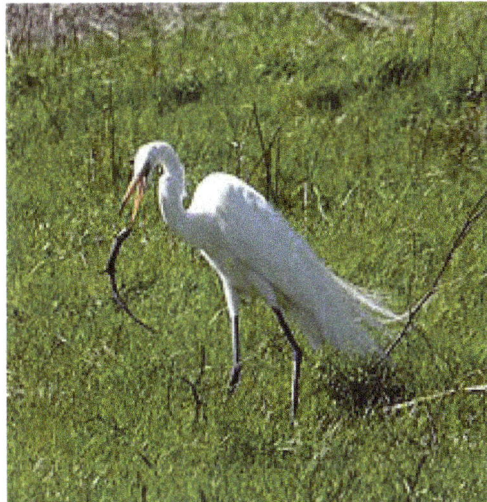

A great egret manipulating its prey, a lizard, prior to swallowing

Black herons holding wings out to form an umbrella-like canopy under which to hunt

The most common hunting technique is for the bird to sit motionless on the edge of or standing in shallow water and to wait until prey comes within range. Birds may either do this from an upright posture, giving them a wider field of view for seeing prey, or from a crouched position, which is more cryptic and means the bill is closer to the prey when it is located. Having seen prey, the head is moved from side to side, so that the heron can calculate the position of the prey in the water and compensate for refraction, and then the bill is used to spear the prey.

Tricoloured heron fishing, using wings to create shade

In addition to sitting and waiting, herons may feed more actively. They may walk slowly, around or less than 60 paces a minute, snatching prey when it is observed. Other active feeding behaviours include foot stirring and probing, where the feet are used to flush out hidden prey. The wings may be used to frighten prey (or possibly attract it to shade) or to reduce glare; the most extreme example of this is exhibited by the black heron, which forms a full canopy with its wings over its body.

Some species of heron, such as the little egret and grey heron, have been documented using bait to lure prey to within striking distance. Herons may use items already in place, or actively add items to the water to attract fish such as the banded killifish. Items used may be man-made, such as bread; alternatively, striated herons in the Amazon have been watched repeatedly dropping seeds, insects, flowers, and leaves into the water to catch fish.

Three species, the black-headed heron, whistling heron, and especially the cattle egret, are less tied to watery environments and may feed far away from water. Cattle egrets improve their foraging success by following large grazing animals, catching insects flushed by their movement. One study found that the success rate of prey capture increased 3.6 times over solitary foraging.

Breeding

While the family exhibits a range of breeding strategies, overall, the herons are monogamous and mostly colonial. Most day herons and night herons are colonial, or partly colonial depending on circumstances, whereas the bitterns and tiger herons are mostly solitary nesters. Colonies may contain several species, as well as other species of waterbirds. In a study of little egrets and cattle egrets in India, the majority of the colonies surveyed contained both species. Nesting is seasonal in temperate species; in tropical species, it may be seasonal (often coinciding with the rainy season) or year-round. Even in year-round breeders, nesting intensity varies throughout the year. Tropical herons typically have only one breeding season per year, unlike some other tropical birds which may raise up to three broods a year.

The larger bitterns, like this American bittern, are solitary breeders. To advertise for mates, males use loud, characteristic calls, referred to as booming.

Courtship usually takes part on the nest. Males arrive first and begin the building of the nest, where they display to attract females. During courtship, the male employs a stretch display and uses erectile neck feathers; the neck area may swell. The female risks an aggressive attack if she approaches too soon and may have to wait up to four days. In colonial species, displays involve visual cues, which can include adopting postures or ritual displays, whereas in solitary species, auditory cues, such as the deep booming of the bitterns, are important. The exception to this is the boat-billed heron, which pairs up away from the nesting site. Having paired, they continue to build the nest in almost all species, although in the little bittern and least bittern, only the male works on the nest.

Some ornithologists have reported observing female herons attaching themselves to impotent mates, then seeking sexual gratification elsewhere.

The nests of herons are usually found near or above water. They are typically placed in vegetation, although the nests of a few species have been found on the ground where suitable trees of shrubs are unavailable. Trees are used by many species, and here they may be placed high up from the ground, whereas species living in reed beds may nest very close to the ground.

Generally, herons lay between three and seven eggs. Larger clutches are reported in the smaller bitterns and more rarely some of the larger day herons, and single-egg clutches are reported for some of the tiger herons. Clutch size varies by latitude within species, with individuals in temperate climates laying more eggs than tropical ones. On the whole, the eggs are glossy blue or white, with the exception being the large bitterns, which lay olive brown eggs.

Name

The word "heron" is rather old. It appeared in the English language around 1300, originating from Old French *hairon, eron* (12th century), earlier *hairo* (11th century), from Frankish *haigiro* or from Proto-Germanic **haigrô, *hraigrô*.

Herons are also known as "shitepokes" or euphemistically as "shikepokes" or "shypokes". *Webster's Dictionary* suggests that herons were given this name because of their habit of defecating when flushed. The terms "shitepoke" or "shikepoke" can be used as insults in a number of situations. For example, the term "shikepoke" appears in the 1931 play *Green Grow the Lilacs*, and in the 1943 musical play *Oklahoma!*.

The 1971 *Compact Edition of the Oxford English Dictionary* describes the use of "shitepoke" for the small green heron of North America (*Butorides virescens*) as originating in the United States, citing a published example from 1853. The *OED* also observes that "shiterow" or "shederow" are terms used for herons, and also applied as derogatory terms meaning a "thin, weakly person". This name for a heron is found in a list of gamebirds in a royal decree of James VI (1566–1625) of Scotland. The *OED* speculates that "shiterow" is a corruption of "shiteheron".

Another former name was "heronshaw". Corrupted to "handsaw", this name appears in Shakespeare's *Hamlet*. A possible further corruption took place in the Norfolk Broads, where the heron is often referred to as a "harnser".

Taxonomy and Systematics

Analyses of the skeleton, mainly the skull, suggested that the Ardeidae could be split into a diurnal and a crepuscular/nocturnal group which included the bitterns. From DNA studies and skeletal analyses focusing more on bones of body and limbs, this grouping has been revealed as incorrect. Rather, the similarities in skull morphology reflect convergent evolution to cope with the different challenges of daytime and night-time feeding. Today, it is believed that three major groups can be distinguished, which are (from the most primitive to the most advanced):

- tiger herons and the boatbill

- bitterns

- day herons and egrets, and night herons

The night herons could warrant separation as subfamily Nycticoracinae, as it was traditionally done. However, the position of some genera (e.g. *Butorides* or *Syrigma*) is unclear at the moment, and molecular studies have until now suffered from a small number of studied taxa. Especially, the relationships among the Ardeinae subfamily are very badly resolved. The arrangement presented here should be considered provisional.

A 2008 study suggests that this family belongs to the Pelecaniformes. In response to these findings, the International Ornithological Congress recently reclassified Ardeidae and their sister taxa Threskiornithidae under the order Pelecaniformes instead of the previous order of Ciconiiformes.

Great bittern (*Botaurus stellaris*)

Eastern great egret (*Ardea modesta*)

The wounded heron by George Frederic Watts, 1837 (Watts Gallery)

Subfamily Tigriornithinae

- Genus *Cochlearius* – boat-billed heron

- Genus *Tigrisoma* – typical tiger herons (three species)

- Genus *Tigriornis* – white-crested tiger heron

- Genus *Zonerodius* – forest bittern

Subfamily Botaurinae

- Genus *Zebrilus* – zigzag heron

- Genus *Ixobrychus* – small bitterns (eight living species, one recently extinct)

- Genus *Botaurus* – large bitterns (four species)

Subfamily Ardeinae

- Genus *Zeltornis* (fossil)

- Genus *Nycticorax* – typical night herons (two living species, four recently extinct; includes *Nyctanassa*)

- Genus *Nyctanassa* – American night herons (one living species, one recently extinct)

- Genus *Gorsachius* – Asian and African night herons (four species)

- Genus *Butorides* – green-backed herons (three species; sometimes included in *Ardea*)

- Genus *Agamia* – Agami heron

- Genus *Pilherodius* – capped heron

- Genus *Ardeola* – pond herons (six species)

- Genus *Bubulcus* – cattle egrets (one or two species, sometimes included in *Ardea*)

- Genus *Proardea* (fossil)

- Genus *Ardea* – typical herons (11–17 species)

- Genus *Syrigma* – whistling heron

- Genus *Egretta* – typical egrets (seven–13 species)

- Genus undetermined

  o   Easter Island heron, Ardeidae *gen. et sp. indet.* (prehistoric)

Fossil Herons of Unresolved Affiliations

- *Calcardea* – (Paleocene)

- *Xenerodiops* – (Early Oligocene of Fayyum, Egypt)

- *"Anas" basaltica* –(Late Oligocene of Varnsdorf, Czech Republic)

- *Ardeagradis*

- *Proardeola* – possibly same as *Proardea*

- *Matuku otagoense* – (Early Miocene of Otago, New Zealand)

Other prehistoric and fossil species are included in the respective genus accounts. In addition, *Proherodius* is a disputed fossil which was variously considered a heron or one of the extinct long-legged waterfowl, the Presbyornithidae. It is only known from a sternum; a tarsometatarsus assigned to it actually belongs to the paleognath *Lithornis vulturinus*.

Grebe

A grebe is a member of the order Podicipediformes and the only type of bird associated with this order.

Grebes are a widely distributed order of freshwater diving birds, some of which visit the sea when migrating and in winter. This order contains only a single family, the Podicipedidae, containing 22 species in 6 extant genera.

Description

Diving grebe

Grebes are small to medium-large in size, have lobed toes, and are excellent swimmers and divers. Although they can run for a short distance, they are prone to falling over, since they have their feet placed far back on the body.

Grebes have narrow wings, and some species are reluctant to fly; indeed, two South American species are completely flightless. They respond to danger by diving rather than flying, and are in any case much less wary than ducks. Extant species range in size from the least grebe, at 120 grams (4.3 oz) and 23.5 cm (9.3 inches), to the great grebe, at 1.7 kg (3.8 lbs) and 71 cm (28 inches).

The North American and Eurasian species are all, of necessity, migratory over much or all of their ranges, and those species that winter at sea are also seen regularly in flight. Even the small freshwater pied-billed grebe of North America has occurred as a transatlantic vagrant to Europe on more than 30 occasions.

Bills vary from short and thick to long and pointed, depending on the diet, which ranges from fish to freshwater insects and crustaceans. The feet are always large, with broad lobes on the toes and small webs connecting the front three toes. The hind toe also has a small lobe. Recent experimental work has shown that these lobes work like the hydrofoil blades of a propeller. Curiously, the same mechanism apparently evolved independently in the extinct Cretaceous-age Hesperornithiformes, which are totally unrelated birds.

Grebes have unusual plumage. It is dense and waterproof, and on the underside the feathers are at right-angles to the skin, sticking straight out to begin with and curling at the tip. By pressing their feathers against the body, grebes can adjust their buoyancy. Often, they swim low in the water with just the head and neck exposed.

In the non-breeding season, grebes are plain-coloured in dark browns and whites. However, most have ornate and distinctive breeding plumages, often developing chestnut markings on the head area, and perform elaborate display rituals. The young, particularly those of the *Podiceps* genus, are often striped and retain some of their juvenile plumage even after reaching full size. In the breeding season, they mate at freshwater lakes and ponds, but some species spend their non-breeding season along seacoasts.

When preening, grebes eat their own feathers, and feed them to their young. The function of this behaviour is uncertain but it is believed to assist with pellet formation, and to reduce their vulnerability to gastric parasites.

Grebes make floating nests of plant material concealed among reeds on the surface of the water. The young are precocial, and able to swim from birth.

Taxonomy, Systematics and Evolution

The grebes are a radically distinct group of birds as regards their anatomy. Accordingly, they were at first believed to be related to the loons, which are also foot-propelled diving birds, and both families were once classified together under the order Colymbiformes. However, as recently as the 1930s, this was determined to be an example of convergent evolution by the strong selective forces encountered by unrelated birds sharing the same lifestyle at different times and in different habitat. Grebes and loons are now separately classified orders of Podicipediformes and Gaviiformes, respectively.

The cladistics vs. phenetics debate of the mid-20th century revived scientific interest in generalizing comparisons. As a consequence, the discredited grebe-loon link was discussed again. This even went as far as proposing monophyly for grebes, loons, and the toothed Hesperornithiformes. In retrospect, the scientific value of the debate lies more in providing examples that a cladistic *methodology* is not incompatible with an overall phenetical scientific *doctrine*, and that thus, simply because some study "uses cladistics", it does not guarantee superior results.

Molecular studies such as DNA-DNA hybridization (Sibley & Ahlquist, 1990) and sequence analyses fail to resolve the relationships of grebes properly due to insufficient resolution in the former and long-branch attraction in the latter. Still – actually *because* of this – they do confirm that these birds form a fairly ancient evolutionary lineage (or possibly one that was subject to selective pressures down to the molecular level even), and they support the *non*-relatedness of loons and grebes.

The most comprehensive study of bird phylogenomics, published in 2014, found that grebes and flamingos are members of Columbea, a clade that also includes doves, sandgrouse, and mesites.

Relationship with Flamingos

Many molecular and morphological studies support a relationship between grebes and flamingos.

Recent molecular studies have suggested a relation with flamingos while morphological evidence also strongly supports a relationship between flamingos and grebes. They hold at least eleven morphological traits in common, which are not found on other birds. Many of these characteristics have been previously identified on flamingos, but not on grebes. The fossil Palaelodids can be considered evolutionarily, and ecologically, intermediate between flamingos and grebes.

For the grebe-flamingo clade, the taxon Mirandornithes ("miraculous birds" due to their extreme divergence and apomorphies) has been proposed. Alternatively, they could be placed in one order, with Phoenocopteriformes taking priority.

Fossil Grebes

The fossil record of grebes is incomplete; there are no transitional forms between more conventional birds and the highly derived grebes known from fossils, or at least none that can be placed in the relationships of the group with any certainty. The enigmatic waterbird genus *Juncitarsus*, however, may be close to a common ancestor of flamingos and grebes.

The Early Cretaceous (Berriasian, around 143 mya) genus *Eurolimnornis* from NW Romania was initially believed to be a grebe. If it is indeed related to this lineage, it must represent a most basal form, as it almost certainly predates any grebe-flamingo split. On the other hand, the single bone fragment assigned to this taxon is not very diagnostic and may not be of a bird at all.

Telmatornis from the Navesink Formation – also Late Cretaceous – is traditionally allied with the Charadriiformes and/or Gruiformes. However, a cladistic analysis of the forelimb skeleton found it highly similar to the great crested grebe and unlike the painted buttonquail (now known to be a basal charadriiform lineage), the black-necked stilt (a more advanced charadriiform), or the limpkin (a member of the Grui suborder of Gruiformes), namely in that its dorsal condyle of the humerus was not angled at 20°–30° away from long axis of the humerus. The analysis did not result in a phylogenetic pattern but rather grouped some birds with similar wing shapes together while others stood separate. It is thus unknown whether this apparent similarity to grebes represents an evolutionary relationship, or whether *Telmatornis* simply had a wing similar to that of grebes and moved it like they do.

Juncitarsus merkeli fossil

True grebes suddenly appear in the fossil record in the Late Oligocene or Early Miocene, around 23–25 mya. While there are a few prehistoric genera that are now completely extinct; *Thiornis* (Late Miocene -? Early Pliocene of Libros, Spain) and *Pliolymbus* (Late Pliocene of WC USA – Early? Pleistocene of Chapala, Mexico) date from a time when most if not all extant genera were already present. Only the Early Miocene *Miobaptus* from Czechoslovakia might be somewhat closer to the ancestral grebes, but more probably belongs to an extinct lineage. Indeed, *Miobaptus* is rivalled or even exceeded in age by a species of the modern genus *Podiceps*.

A few more recent grebe fossils could not be assigned to modern or prehistoric genera:

- Podicipedidae gen. et sp. indet. (San Diego Late Pliocene of California) – formerly included in *Podiceps parvus*

- Podicipedidae gen. et sp. indet. UMMP 49592, 52261, 51848, 52276, KUVP 4484 (Late Pliocene of WC USA)

- Podicipedidae gen. et sp. indet. (Glenns Ferry Late Pliocene/Early Pleistocene of Idaho, USA)

Grebes date back very far and the Late Cretaceous bird *Neogaeornis wetzeli* may be their ancestor.

River Dolphin

River dolphins are a widely distributed group of fully aquatic mammals that reside exclusively in freshwater or brackish water. They are an informal grouping of dolphins, which is a paraphyletic group within the infraorder Cetacea. The river dolphins comprise the extant families Platanistidae (the Indian dolphins), Iniidae (the Amazonian dolphins), and Pontoporiidae (the brackish dolphins). There are five extant species of river dolphins, and two subspecies. River dolphins, alongside other cetaceans, belong to the clade Cetartiodactyla, with even-toed ungulates, and their closest living relatives the hippopotamuses, having diverged about 40 million years ago.

River dolphins are relatively small compared to other dolphins, having evolved to survive in warm, shallow water and strong river currents. They range in size from the 5-foot (1.5 m) long South Asian river dolphin to the 8-foot (2.4 m) and 220-pound (100 kg) Amazon river dolphin. Several species exhibit sexual dimorphism, in that the males are larger than the females. They have streamlined bodies and two limbs that are modified into flippers. River dolphins use their conical-shaped teeth and long beaks to capture fast-moving prey in murky water. They have well-developed hearing that is adapted for both air and water; they do not really rely on vision since the water they swim in is usually very muddy. These species are well-adapted to living in warm, shallow waters, and, unlike other cetaceans, have little to no blubber.

River dolphins are not very widespread; they are all restricted to certain rivers or deltas. This makes them extremely vulnerable to habitat destruction. River dolphins feed primarily on fish. Male river dolphins typically mate with multiple females every year, but females only mate every two to three years. Calves are typically born in the spring and summer months and females bear all the responsibility for raising them. River dolphins produce a variety of vocalizations, usually in the form of clicks and whistles.

River dolphins are rarely kept in captivity; breeding success has been poor and the animals often die within a few months of capture. As of 2015, there are only three river dolphins in captivity.

Taxonomy and Evolution

Classification

Four families of river dolphins (Iniidae, Pontoporiidae, Lipotidae and Platanistidae) are currently recognized, comprising three superfamilies (Inioidea, Lipotoidea and Platanistoidea). Platanistidae, containing the two subspecies of South Asian river dolphin, is the only accepted family of Platanistoidea. Previously, many taxonomists had assigned all river dolphins to a single family, Platanistidae, and treated the Ganges and Indus River dolphins as separate species. A December 2006 survey found no members of *Lipotes vexillifer* (commonly known as the baiji, or Chinese river dolphin) and declared the species functionally extinct. With their disappearance, one of the recently accepted superfamilies, Lipotoidea, has become extinct.

The current classification of river dolphins is as follows:

Life reconstruction of *Arktocara yakataga*, an Allodelphinidae

In 2012 the Society for Marine Mammalogy began considering the Bolivian (*Inia geoffrensis boliviensis*) and Amazonian (*Inia geoffrensis geoffrensis*) subspecies as full species *Inia boliviensis* and *Inia geoffrensis*, respectively; however, much of the scientific community, including the IUCN, continue to consider the Bolivian population to be a subspecies of *Inia geoffrensis*.

In October 2014, the Society for Marine Mammalogy took *Inia boliviensis* and *Inia araguaiaensis* off their list of aquatic mammal species and subspecies and currently does not recognize these species-level separations.

Evolution

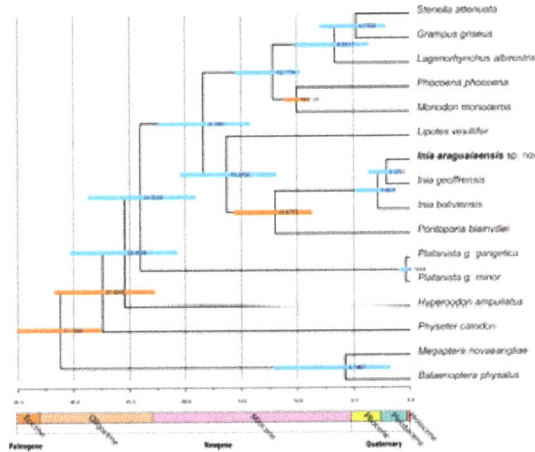

Phylogeny of cetaceans based on cytochrome b gene sequences, showing the distant relationship between *Platanista* and other river dolphins.

River dolphins are members of the infraorder *Cetacea*, which are descendants of land-dwelling mammals of the order Artiodactyla (even-toed ungulates). They are related to the *Indohyus*, an extinct chevrotain-like ungulate, from which they split approximately 48 million years ago. The primitive cetaceans, or archaeocetes, first took to the sea approximately 49 million years ago and became fully aquatic by 5–10 million years later. It is unknown when river dolphins first ventured back into fresh water.

River dolphins are thought to have relictual distributions, that is, their ancestors originally occupied marine habitats, but were then displaced from these habitats by modern dolphin lineages. Many of the morphological similarities and adaptations to freshwater habitats arose due to convergent evolution; thus, a grouping of all river dolphins is paraphyletic. Amazon river dolphins are actually more closely related to oceanic dolphins than to South Asian river dolphins. *Isthminia panamensis* is an extinct genus and species of river dolphin, living 5.8 to 6.1 million years ago. Its fossils were discovered near Piña, Panama.

River dolphin has been considered a taxonomic description, suggesting an evolutionary relationship among the group, although it is now known that they form two distinct clades. 'True' river dolphins are descendants of ancient evolutionary lineages that evolved in freshwater environments.

Some species of cetacean live in rivers and lakes, but are more closely related to oceanic dolphins or porpoises and entered fresh water more recently. Such species are considered facultative freshwater cetaceans as they can use both marine and freshwater environments. These include species such as the Irrawaddy dolphin, *Orcaella brevirostris*, found in the Mekong, Mahakam, the Irrawaddy Rivers, as well as the Yangtze finless porpoise *Neophocaena phocaenoides asiaeorientalis*.

The tucuxi (*Sotalia fluviatilis*) in the Amazon River is another species descended from oceanic dolphins; however, it does not perfectly fit the label of 'facultative' either, as it occurs only in fresh water. The tucuxi was until recently considered conspecific with the Guiana dolphin (*Sotalia guianensis*), which inhabits marine waters. It may also be true for the Irrawaddy dolphin and the finless porpoise that, although the species may be found in both freshwater and marine environments, individual animals found in rivers may not be able to survive in the ocean, and vice versa. The tucuxi is currently classified as an oceanic dolphin (Delphinidae).

The Franciscana (*Pontoporia blainvillei*) has shown a converse evolutionary pattern, and has an ancient evolutionary lineage in freshwater, but inhabits estuarine and coastal waters.

Biology

Anatomy

River dolphins have a torpedo shaped body with a flexible neck, limbs modified into flippers, non-existent external ear flaps, a tail fin, and a small bulbous head. River dolphin skulls have small eye orbits, a long snout and eyes placed on the sides of the head. River dolphins are rather small, ranging in size from the 5-foot (1.5 m) long South Asian river dolphin to the 8-foot (2.4 m) and 220-pound (100 kg) Amazon river dolphin. They all have female-biased sexual dimorphism, with the females being larger than the males. River dolphins are polygynous, meaning male river dolphins typically mate with multiple females every year, but females only mate every two to three years. Calves are typically born in the spring and summer months and females bear all the responsibility for raising them.

Indus river dolphin skull

River dolphins have conical teeth, used to catch swift prey such as small river fish. They also have very long snouts, with some measuring 23 inches (58 cm), four times longer than most of their oceanic counterparts. They have a two-chambered stomach that is similar in structure to that of terrestrial carnivores. They have fundic and pyloric chambers. Breathing involves expelling stale air from their blowhole, followed by inhaling fresh air into their lungs. They do not have the iconic spout, as this only forms when the warm air exhaled from the lungs meets cold external air, which does not occur in their tropical habitats.

River dolphins have a relatively thin layer of blubber. Blubber can help with buoyancy, protection from predators (they would have a hard time getting through a thick layer of fat), energy for leaner times, and insulation from harsh climates. The habitats of river dolphins lack these needs.

Locomotion

River dolphins have two flippers and a tail fin. These flippers contain four digits. Although river dolphins do not possess fully developed hind limbs, some possess discrete rudimentary appendages, which may contain feet and digits. River dolphins are slow swimmers in comparison to oceanic dolphins, which can travel at speeds up to 35 miles per hour (56 km/h); the tucuxi can only travel at about 14 miles per hour (23 km/h). Unlike other cetaceans, their neck vertebrae are not fused together, meaning they have greater flexibility than other non-terrestrial aquatic mammals, at the expense of speed. This means they can turn their head without actually moving their entire body. When swimming, river dolphins rely on their tail fins to propel themselves through the water. Flipper movement is continuous. River dolphins swim by moving their tail fins and lower bodies up and down, propelling themselves through vertical movement, while their flippers are mainly used for steering. All species have a dorsal fin.

Senses

Biosonar by cetaceans

The ears of river dolphins have specific adaptations to their aquatic environment. In humans, the middle ear works as an impedance equalizer between the outside air's low impedance and the cochlear fluid's high impedance. In river dolphins, and other cetaceans, there is no great difference between the outer and inner environments. Instead of sound passing through the outer ear to the middle ear, river dolphins receive sound through the throat, from which it passes through a low-impedance fat-filled cavity to the inner ear. The ear is acoustically isolated from the skull by air-filled sinus pockets, which allows for greater directional hearing underwater. Dolphins send out high frequency clicks from an organ known as a melon. This melon consists of fat, and the skull of any such creature containing a melon will have a large depression. This allows river dolphins to produce biosonar for orientation. They are so dependent on echolocation that they can survive even if they are blind. Beyond locating an object, echolocation also provides the animal with an idea on the object's shape and size, though how exactly this works is not yet understood. The small hairs on the rostrum of the Amazon river dolphin are believed to function as a tactile sense, possibly to compensate for their poor eyesight.

River dolphins have very small eyes

River dolphins have very small eyes for their size, and do not have a very good sense of sight. In addition, the eyes are placed on the sides of the head, so the vision consists of two fields, rather than a binocular view like humans have. When river dolphins surface, their lens and cornea correct the nearsightedness that results from the refraction of light. They have both rod and cone cells, meaning they can see in both dim and bright light. Most river dolphins have slightly flattened eyeballs, enlarged pupils (which shrink as they surface to prevent damage), slightly flattened corneas and a tapetum lucidum; these adaptations allow for large amounts of light to pass through the eye and, therefore, a very clear image of the surrounding area. They also have glands on their eyelids and an outer corneal layer that act as protection for the cornea.

Olfactory lobes are absent in river dolphins, suggesting that they have no sense of smell.

River dolphins are not thought to have a sense of taste, as their taste buds are atrophied or missing altogether. However, some dolphins have preferences between different kinds of fish, indicating some sort of attachment to taste.

Interactions with Humans

Development

Development and agriculture have had devastating impacts on the habitats on river dolphins. The total population of Araguaian river dolphins is estimated to be between 600 and 1,500 individuals, and genetic diversity is limited. The ecology of their habitat has been adversely affected by agricultural, ranching and industrial activities, as well as by the use of dams for hydroelectric power. The inhabited section of the Araguaia River probably extends over about 900 miles (1,400 km) out of a total length of 1,300 miles (2,100 km). The Tocantins river habitat is fragmented by six hydroelectric dams, so the population there is at particular risk. Its probable eventual IUCN status is Vulnerable or worse.

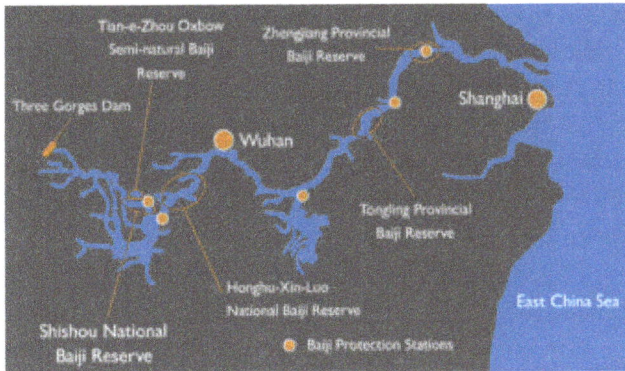

Conservation areas for the baiji along the Yangtze River

Both subspecies of South Asian river dolphins have been very adversely affected by human use of the river systems in the subcontinent. Irrigation has lowered water levels throughout both subspecies' ranges. Poisoning of the water supply from industrial and agricultural chemicals may have also contributed to population decline. Perhaps the most significant issue is the building of more than 50 dams along many rivers, causing the segregation of populations and a narrowed gene pool in which the dolphins can breed. Currently, three subpopulations of Indus river dolphins are considered capable of long-term survival if protected.

As China developed economically, pressure on the baiji river dolphin grew significantly. Industrial and residential waste flowed into the Yangtze. The riverbed was dredged and reinforced with concrete in many locations. Ship traffic multiplied, boats grew in size, and fishermen employed wider and more lethal nets. Noise pollution caused the nearly blind animal to collide with propellers. Stocks of the dolphin's prey declined drastically in the late 20th century, with some fish populations declining to one thousandth of their pre-industrial levels. In the 1950s, the population was estimated at 6,000 animals, but declined rapidly over the subsequent five decades. Only a few hundred were left by 1970. Then the number dropped down to 400 by the 1980s and then to 13 in 1997 when a full-fledged search was conducted. On December 13, 2006, the baiji (*Lipotes vexillifer*) was declared "functionally extinct", after a 45-day search by leading experts in the field failed to find a single specimen. The last verified sighting was in September 2004.

Competition

The region of the Amazon in Brazil has an extension of 3,100,000 sq mi (8,000,000 km²) containing diverse fundamental ecosystems. One of these ecosystems is a floodplain, or a várzea forest, and is home to a large number of fish species which are an essential resource for human consumption. The várzea is also a major source of income through excessive local commercialized fishing. Várzea consist of muddy river waters containing a vast number and diversity of nutrient-rich species. The abundance of distinct fish species lures the Amazon River dolphin into the várzea areas of high water occurrences during the seasonal flooding.

In addition to attracting predators such as the Amazon river dolphin, these high-water occurrences are an ideal location to draw in the local fisheries. Human fishing activities directly compete with the dolphins for the same fish species, the tambaqui (*Colossoma macropomum*) and the pirapitinga (*Piaractus brachypomus*), resulting in deliberate or unintentional catches of the Amazon river dolphin. The local fishermen overfish, and when the Amazon River dolphins remove the commercialized fish from the nets and lines, it damages the equipment and the capture and causes a negative reaction from the local fishermen. The Brazilian Institute of Environment and Renewable Natural Resources prohibit fishermen from killing the Amazon river dolphin, yet they are not compensated for the damage to their equipment and the loss of their catch.

Bycatch

During the process of catching the commercialized fish, the Amazon river dolphins get caught in the nets and exhaust themselves until they die, or the local fishermen deliberately kill the dolphins that become entangled in their nets. The carcasses are discarded, consumed, or used as bait to attract a scavenger catfish, the piracatinga (*Calophysus macropterus*). The use of the Amazon river dolphin carcass as bait for the piracatinga dates back from 2000. The increasing consumption demand by the local inhabitants and Colombia for the piracatinga has created a market for distribution of the Amazon river dolphin carcasses to be used as bait throughout these regions.

For example, of the 15 dolphin carcasses found in the Japurá River in 2010–2011 surveys, 73% of the dolphins were killed for bait, disposed of, or abandoned in entangled gillnets. The data does not fully represent the actual overall number of deaths of the Amazon river dolphins, whether accidental or intentional, because a variety of factors make it extremely complicated to record and medically examine all the carcasses. Scavenger species feed upon them and the complexity of the river currents makes it nearly impossible to locate all the carcasses. More importantly, the local fishermen do not report these deaths out of fear that legal action will be taken against them, as the Amazon river dolphin and other cetaceans are protected under the Brazilian federal law, prohibiting any takes, harassments, and kills of the species.

In Captivity

A baiji conservation dolphinarium was established at the Institute of Hydrobiology (IHB) in Wuhan in 1992. This was planned as a backup to any other conservation efforts by producing an area completely protected from any threats, and where the baiji could be easily observed. The site includes an indoor and outdoor holding pool, a water filtration system, food storage and preparation facilities, research labs and a small museum. The aim is to also generate income from tourism which can be put towards the baiji plight. The pools are not very large, only kidney shaped tanks with dimensions of 82 feet (25 m) arc 23 feet (7.0 m) width and 11 feet (3.4 m) depth, 33 feet (10 m) diameter, 6.6 feet (2.0 m) deep and 39 feet (12 m) diameter, 11 feet (3.4 m) deep, and are not capable of holding many baijis at one time. Douglas Adams and Mark Carwardine documented their encounters with the endangered animals on their conservation travels for the BBC programme *Last Chance to See*. The book by the same name, published in 1990, included pictures of a captive specimen, a male named Qi Qi that lived in the Wuhan Institute of Hydrobiology dolphinarium from 1980 to July 14, 2002. Discovered by a fisherman in Dongting Lake, he became the sole resident of the Baiji Dolphinarium beside East Lake. A sexually mature female was captured in late 1995, but died after half a year in 1996 when the Shishou Tian-e-Zhou Baiji Semi-natural Reserve, which had contained only finless porpoises since 1990, was flooded.

The only trained Amazon river dolphin in the world at the Acuario de Valencia, Venezuela

The Amazon river dolphin has historically been kept in dolphinariums. Today, only three exist in captivity: one in Acuario de Valencia in Venezuela, one in Zoologico de Guistochoca in Peru, and one in Duisburg Zoo in Germany. Several hundred were captured between the 1950s and 1970s, and were distributed in dolphinariums throughout the US, Europe, and Japan. Around 100 went to US dolphinariums, and of that, only 20 survived; the last (named Chuckles) died in Pittsburgh Zoo in 2002.

In Mythology

Ganga on a river dolphin

Old World

In Hindu mythology, the Ganges River Dolphin is associated with Ganga, the deity of the Ganges river. The dolphin is said to be among the creatures which heralded the goddess' descent from the heavens and her mount, the Makara, is sometimes depicted as a dolphin.

In Chinese mythology, the baiji has many origin stories. For example, near the mouth of the Yangtze, the baiji was a princess that had lost her parents and had lived with her step-father, whom she had longed to get away from. The step-father wanted to trade her since she would have been sold for a great sum of money, but as they were crossing the river to get to the trader, a storm rolled in, and they were drenched. The step-father, enraged, tried to take her, but she plunged herself into the river. Before being drowned in the river, she was transformed into a dolphin, and swam away from her abusive step-father, who also fell in and was transformed into a porpoise.

In another story further upstream the Yangtze, the baiji was the daughter of a general who was deported from the city of Wuhan during a war. During his duty, the daughter ran away. Later, the general met a woman who told him how her father was a general, and when he realized that she was his daughter, he threw himself into the river out of shame. The daughter ran after him and also fell into the river. Before they were drowned, the daughter was transformed into a dolphin, and the general a porpoise.

New World

Amazon river dolphins, known by the natives as the boto or encantados, are very prevalent in the mythology of the native South Americans. They are often characterized in their mythology as wielding superior musical ability, their seductiveness and love of sex, often resulting in illegitimate children, and their attraction to parties. Despite the fact that the Encante are said to come from a utopia full of wealth and without pain or death, the encantados crave the pleasures and hardships of human societies.

Transformation into human form is said to be rare, and usually occurs at night. The encantado will often be seen running from a festa, despite protests from the others for it to stay, and can be seen by pursuers as it hurries to the river and reverts to dolphin form. When it is under human form, it wears a hat to hide its blowhole, which does not disappear with the shapeshift.

Besides the ability to shapeshift into human form, encantados frequently wield other magical abilities, such as the power to control storms, enchant humans into doing their will, transform humans into encantados themselves, and inflict illness, insanity, and even death. Shamans often intervene in these situations.

Kidnapping is also a common theme in such folklore. Encantados are said to be fond of abducting humans with whom they fall in love, children born of their illicit love affairs, or just about anyone near the river who can keep them company, and taking them back to the

Encante. The fear of this is so great among people who live in the Amazon river area that both children and adults are terrified of going near the water between dusk and dawn, or entering water-bodies alone. Some who supposedly have encountered encantados while out in their canoes have been said to have gone insane, although in fact, the creatures seem to have done little more than follow their boats and nudge them from time to time.

References

- Jell, Peter A. (1980). "Earliest known pelecypod on Earth — a new Early Cambrian genus from South Australia". Alcheringa: An Australasian Journal of Palaeontology. 4 (3): 233–239. doi:10.1080/03115518008618934

- Dorit, Robert L.; Walker, Warren F. Jr.; Barnes, Robert D. (1991). Zoology. Saunders College Publishing. p. 674. ISBN 978-0-03-030504-7

- Wells, Roger M. (1998). "Class Bivalvia". Invertebrate Paleontology Tutorial. State University of New York College at Cortland. Retrieved 2012-04-11

- Kennedy, W. J.; Taylor, J. D.; Hall, A. (1969). "Environmental and biological controls on bivalve shell mineralogy". Biological Reviews. 44 (4): 499–530. doi:10.1111/j.1469-185X.1969.tb00610.x

- Hall, K. R. L.; Schaller, G. B. (1964). "Tool-using behavior of the California sea otter". Journal of Mammalogy. 45 (2): 287–298. JSTOR 1376994. doi:10.2307/1376994

- Piper, Ross (2007). Extraordinary Animals: An Encyclopedia of Curious and Unusual Animals. Greenwood Press. pp. 224–225. ISBN 978-0-313-33922-6

- Helm, M. M.; Bourne, N.; Lovatelli, A. (2004). "Gonadal development and spawning". Hatchery culture of bivalves: a practical manual. FAO. Retrieved 2012-05-08

- Morton, B. (2008). "The evolution of eyes in the Bivalvia: new insights". American Malacological Bulletin. 26 (1–2): 35–45. doi:10.4003/006.026.0205

- Wekell, John C.; Hurst, John; Lefebvre, Kathi A. (2004). "The origin of the regulatory limits for PSP and ASP toxins in shellfish" (PDF). Journal of Shellfish Research. 23 (3): 927–930

- Barnes, R. S. K.; Callow, P.; Olive, P. J. W. (1988). The Invertebrates: A New Synthesis. Blackwell Scientific Publications. pp. 132–134. ISBN 978-0-632-03125-2

- Naylor, Martin (2005). "American jack knife clam, (Ensis directus)" (PDF). Alien species in Swedish seas and coastal areas. Retrieved 2012-04-18

- Colicchia, G.; Waltner, C.; Hopf, M.; Wiesner, H. (2009). "The scallop's eye—a concave mirror in the context of biology". Physics Education. 44 (2): 175–179. Bibcode:2009PhyEd..44..175C. doi:10.1088/0031-9120/44/2/009

- Bieler, R.; Mikkelsen, P. M. (2006). "Bivalvia – a look at the branches». Zoological Journal of the Linnean Society. 148 (3): 223–235. doi:10.1111/j.1096-3642.2006.00255.x

- Ponder, W. F.; Lindberg, David R. (2008). Phylogeny and Evolution of the Mollusca. University of California Press. p. 117. ISBN 978-0-520-25092-5

- Reilly, Michael (2009-04-27). "Sea Shells Used to Clean Up Heavy Metals". Discovery News. Archived from the original on March 31, 2012. Retrieved 2012-05-18

- Strong E. E. & Glaubrecht M. (2010). "Anatomy of the Tiphobiini from Lake Tanganyika (Cerithioidea, Paludomidae)". Malacologia 52(1): 115-153. doi:10.4002/040.052.0108

Permissions

All chapters in this book are published with permission under the Creative Commons Attribution Share Alike License or equivalent. Every chapter published in this book has been scrutinized by our experts. Their significance has been extensively debated. The topics covered herein carry significant information for a comprehensive understanding. They may even be implemented as practical applications or may be referred to as a beginning point for further studies.

We would like to thank the editorial team for lending their expertise to make the book truly unique. They have played a crucial role in the development of this book. Without their invaluable contributions this book wouldn't have been possible. They have made vital efforts to compile up to date information on the varied aspects of this subject to make this book a valuable addition to the collection of many professionals and students.

This book was conceptualized with the vision of imparting up-to-date and integrated information in this field. To ensure the same, a matchless editorial board was set up. Every individual on the board went through rigorous rounds of assessment to prove their worth. After which they invested a large part of their time researching and compiling the most relevant data for our readers.

The editorial board has been involved in producing this book since its inception. They have spent rigorous hours researching and exploring the diverse topics which have resulted in the successful publishing of this book. They have passed on their knowledge of decades through this book. To expedite this challenging task, the publisher supported the team at every step. A small team of assistant editors was also appointed to further simplify the editing procedure and attain best results for the readers.

Apart from the editorial board, the designing team has also invested a significant amount of their time in understanding the subject and creating the most relevant covers. They scrutinized every image to scout for the most suitable representation of the subject and create an appropriate cover for the book.

The publishing team has been an ardent support to the editorial, designing and production team. Their endless efforts to recruit the best for this project, has resulted in the accomplishment of this book. They are a veteran in the field of academics and their pool of knowledge is as vast as their experience in printing. Their expertise and guidance has proved useful at every step. Their uncompromising quality standards have made this book an exceptional effort. Their encouragement from time to time has been an inspiration for everyone.

The publisher and the editorial board hope that this book will prove to be a valuable piece of knowledge for students, practitioners and scholars across the globe.

Index